Knowing
Your Intuitive Mind

Advanced
Pendulum Instruction
& Applications

by Dale W. Olson

Volume 1

Crystalline Publications
Eugene, Oregon

Crystalline Publications
P.O. Box 2088
Eugene, Oregon 97402

First Printing August, 1991
Manufactured in The United States of America

ISBN #1-879-246-01-5
Library of Congress Catalog Card # 90-837111

Printed with soy based ink as our part in planetary consciousness.

Dedication

May this book assist you in Knowing your Intuition, the part of you that is connected to the Infinite source of all Love, Light, Knowledge, and Wisdom.

Acknowledgements

Special recognition and thanks to Doctor Marcel Vogel, a great scientist, inventor, healer, and teacher. It has been a great privilege to study under such a wise, loving, and consciously aware human being. He has been a great inspiration in broadening my parameters of knowledge, and a great catalyst in bringing our expanded abilities out of the dark ages through applied science. Personally, his light in healing my heart is beyond words and value, and his help in igniting the light within will be of great value to many.

I give further thanks to Doctor Rudy Zupancic who has been not only a great friend, but also a great influence in showing me that it is the Intuition that separates the health care professionals from the true healers.

I would like to give my limitless thanks and love to Judith Marie Pleskow for all of her great work in bringing this book into form through her endless love, support, and great artistic and publishing talents.

I give special thanks and appreciation to the editor who made this book whole: Hannah Donovan Schmidt, and my proofreader Lee Darling.

Thanks

Note from the author

I am often asked how I became involved in the development of intuitive skills. Over 16 years ago, I found myself entangled in the piles of life's trials. Feeling confused and not knowing the "how to", and the "where to from here", I felt compelled to consult a very wise, elderly "psychic," Johanna Keller. One day, after having had several enlightening sessions with her, she turned to me and said," Dale, you don't need to come to me for the answers, for you have all the answers within you". Of course, I had heard this dozens of times before in numerous books and tapes. Nonetheless, the same old question popped up in my mind, "Yes, but how do you get to that information," or "How do you know when you *know*"? At that moment, Johanna said, "Dale, I'm going to give you two tools to help you find the answers for yourself; the Pendulum and a three-card method for verification". This was the beginning of my work with the Intuition: externalizing the Intuition beyond my subconscious mind. I felt like I was starting at the elementary level. But, for the first time I was able to open communication lines and have direct, conscious rapport with my Intuition. In the years that followed, these techniques and others proved to be very precise and accurate in helping me to determine numerous unknowns: for myself and for hundreds of others searching for answers to their personal unknowns.

One of the highlights in the development of my intuitive capacities happened over 12 years ago in the Bohemia Mountains of Oregon. This area is my favorite camping grounds, and is rich with a history of gold mining dating back to the Gold Rush era of the early 1900's.

I headed into these mountains with nothing more than desire, a set of dowsing rods, a pendulum, and a belief that I would find gold. At the time, I knew absolutely nothing about prospecting, mineralogy, geology, mining or anything else of the sort. I set out to test my intuitive abilities. To shorten a long and adventurous story, by the end of that summer, I had established two hard-rock mining claims and seven placer mining claims amounting to about 180 acres worth of mining operations. I had located both nugget gold in the river beds, and also gold-bearing quartz veins. Also, I had learned a great deal about prospecting, mineralogy, mining, and geology. The tools of my Intuition were dreams, subtle sensing, the pendulum, and dowsing rods that showed me the *"where"*. My dreams had shown me the idea of mining, and the general area to start. The pendulum assisted me in finding the specific area, depth, the exact place to start digging, and the potential amount of gold bearing ore. Yet, it was *persistence, determination, and trust* that helped me find what I needed to know. I had fulfilled my desire for an adventure, and found I could successfully use my intuitive capacities to find gold.

During my adventure I learned significant lessons around illusion and visions of grandeur. The incredibly hard work of mining helped to put the deluding effects of "gold fever" into proper perspective. I never knew there could be so many forms of iron pyrite (fools gold). As they say, "All that glitters is not gold"; a very valuable lesson to keep in mind when making choices that will have long range impact on your life. I did end up with assay reports indicating beyond any question that I had found gold and silver-bearing ore of valid proportions, but not of a high enough grade to warrant the exorbitant cost of setting up a mining and milling operation. Thus, such an operation would be financially impractical. At that time, my desire said to go for further exploration, but my Intuition was telling me to stop. This great adventure in reality was one of the most dangerous things that I could be doing with my life; working with high explosives, close calls with cave-ins, breathing in quartz dust (which is worse than asbestos), and other dangers. And so my mining operation came to an end out of choice.

The Intuition can be a good indicator of when you are getting away from your life's chosen direction for too long, or when you are subjecting yourself to risk that is not for your highest benefit. It took a great deal of strength to put aside my invested interest, desire, and visions of grandeur to listen to that small inner voice of my Intuition saying, "That was a great adventure that you will never forget, and you found what you were looking for. Now its time to get on with your life's work—and mining for a living is not part of it." My success included the proof that we can use our Intuition to learn about something that we have had no prior knowledge. I also learned that we are able to fine-tune our capacities to detect the subtle energies of metals such as gold and silver that may be 50' to 150' below the surface of the ground. Now that thought is quite mind expanding. I did not fulfill my visions of grandeur or great financial riches; nonetheless, it was a rewarding period in the development of my Intuition and I will always have fond memories of a grand adventure that made the whole experience very rich.

My advanced work with the Intuition as it pertains to the health care profession has been inspired by Doctors Rudy Zupancic, Benoytosh Bhattacharyya, and Marcel Vogel. After working with Rudy Zupancic for years in his healing practice, I learned that effective healing takes highly developed communications with both your Intuition and the patient in order to access the source or causation of the symptoms, dis-ease, or dis-order.

From Benoytosh Bhattacharyya, I learned that it is possible to analyze and treat an individual with a dis-order or dis-ease from long distance.

Dr. Marcel Vogel was truly one of the great scientists, inventors and teachers of our times. His work will be reflected throughout the content of this book: the keys he taught include proper use of breath, an open heart and mind, and positive mental-emotional thoughts to bring about health, happiness, and harmony in one's life.

After having spent years working with and duplicating their results, I have learned how to use the pendulum, muscle testing, and dowsing rods for advanced forms of multi-dimensional analysis of the human body through the use of the Intuition. What this means, simply, is that within every human body there is an innate knowledge of everything that is going on within the body, from a complete analysis of the vitamin and mineral content, to deficiencies of any kind, to the exact location and source of all dis-ease. Whether the questioning be of a physical, mental, emotional, or spiritual nature, it is only a matter of the information being available for access to the conscious mind. All information about ourselves is available to us when we learn the art of fine tuning with subtle-sensing and our developed Intuition.

Doctors Zupancic, Bhattacharyya, Vogel, myself and many others agree that there are four main facets of the human "being:" the physical, mental, emotional, and spiritual, and that all these aspects can be accurately and completely accessed through the Intuition. The magnitude of this work is incredible, and it is definitely worth one's time and determination to become proficient in developing the expanded skills of the mind….the Intuition. Through the development of your Intuition, you will find a ten-fold increase in the quality of your life, resulting in greater fulfillment of your mind's potential and an increase in your personal awareness and self-worth. A developed Intuition can save time and money, and may prevent many of the unnecessary tribulations suffered by those who choose the trial-and-error approach to life. This savings in time and accuracy in problem-solving allows for an increase in your health, wealth, and the pursuit of your dreams.

Introduction

We are now in a new age of not only great technological innovations, but also a time with many consciously aware individuals. Our lives are filled with massive numbers of daily choices and decisions which will have an impact on the future of ourselves and others. If we make our choices and decisions completely from our well developed intellect or the five physical senses, then we are limiting our judgement to basically a trial-and-error, or guessing approach to life.

However, by using the Intuition, we may rise above the trial-and-error approach by acknowledging the pool of infinite knowledge. All we need to do is learn the skills to accurately access the Intuition.

Our Intuition is the liaison, or doorway, to the Infinite Intelligence. It is through this intermediary that we may draw upon the forces of Infinite Intelligence at will. It alone contains the secret process by which mental impulses are modified and changed into their spiritual or etheric equivalent. It alone is the medium through which thought may be directed into prayer, and prayer may be transmitted to the source capable of answering prayer.

Unfortunately, in this day and age, marketing is more effective in making sales than does the actual quality or integrity of a new product or service. The labels on products can not possibly disclose all the information that you, as the consumer, may need to know for your well-being. The lack of information, or even misinformation, on some products could cost a person their health, and possibly, their life. We can no longer depend on just our mental mind to make the correct decisions when there are so many unknowns about these products or services. This "buyer beware" market of both goods and services has made it a necessity to develop a functional Intuition that can accurately and effectively communicate with you as to what is, and what is not, for your highest and greatest good. It is a necessity to develop your intuitive abilities in a practical and tangible sense in order to avoid poor choices which could cost you time, money, health, and a decline in the quality of life.

It is no longer necessary to feel the stress that comes from not knowing whether the product that you hold in your hand is toxic, allergenic, non-effective, mildly useful, good, or highly beneficial. Learning how to externalize the Intuition, and being in conscious communication with the Intuition, can raise us above the trial-and-error (guessing) approach to life. The first step is to acknowledge the pool of infinite knowledge available to all of us, and then to learn the skills necessary to accurately access the Intuition.

Anyone with proper training, who has a fairly good ability to concentrate and who has reasonable control over their emotional and mental stability, can use the pendulum, their personal expanded abilities, and become quite accurate within a relatively short period of time. Remember, we are all born with these abilities of knowing and they manifest in different forms, levels, and degrees. It is a natural ability that you can develop like any other skill, taking only *time* and *practice* to acquire. When you develop the Intuition, you learn how to use your mind's great potential to the fullest extent.

Many of you reading this book may, or may not, have had experience in using the pendulum as a tool to externalize your Intuition. Though some of you may already use a pendulum, there is always room for more knowledge and understanding for effective and accurate results. It may be very helpful for you to disregard much of what you thought you knew about the pendulum and be open to developing a new or different approach to this tool of the Intuition.

The individual ability to do effective pendulum work may vary dramatically. Consequently, we will begin by establishing fundamental pendulum or dowsing form. For the advanced student, please bear with the elementary dialogue. You may benefit from both new technique and from reminders of techniques with which you may already be familiar.

This book contains some general principles to promote your personal growth, as well as specific techniques to enhance your intuitive work. I would like to assure you that your Intuition works perfectly right now. There is nothing that I, or anyone else, could present to you that will "make it work better." However, by learning good pendulum form and doing this program and the exercises involved, you *will* develop your ability to identify, value, and trust your Intuition. With practice, you *will* be able to get in touch with your Intuition when you need it, instead of wondering if it will give you the proper message at the appropriate time with the "hit or miss approach."

I will assist you in developing good pendulum form in learning how to apply this skill of the Intuition for everyday decision-making and intuitive problem-solving.

Table of Contents

Chapter 8

Chapter 9

Chapter 1

The Science and Art of the Pendulum

Tools for the Intuition, including the pendulum, go back to 8000 B.C., and were used by the Hebrews, Egyptians, Chinese, Romans, Greeks, Druids, Hindus, Peruvians, and the American Indians. Even Moses was a skilled dowser, and in the Bible, there are several references to Moses being a water wizard, meaning that he was able to find water for the masses by means of a dowsing staff. The Egyptians were also known for using dowsing rods and pendulums, not only for finding water and gold, but also for divination. The Chinese would call for a Radiesthesist, a person adept at using a pendulum or other tools for sensing the unseen energies around the potential building sites to detect, as they would have said, "The Claw of the Dragon", which probably corresponds to what we now call the harmful rays of a positive-charge vortex or positive-charge ley lines.

Dowsing is the art of using a tool (such as a pendulum or dowsing rod) to access the Intuition to gather information not directly available to the conscious mind. It is the coupling of skills learned by the intellectual mind to access the subconscious mind. Dowsing can be done at a particular site, or from a distance to acquire information about a particular site.

Radiesthesia is the detecting and measuring of an entire spectrum of radiations: these can be mineral, plant, animal, or human. Divining encompasses the art of dowsing and radiesthesia. It includes the possibility of Knowing beyond the limitations of one's five physical senses, including sensing of future results or events, with or without, the use of a device, cards, or symbols.

The goal of dowsing, historically and presently, is to access information not available to our conscious minds, through the use of our Intuition and specific tools. Implicit in this objective is the desire to help someone or oneself, hence it is most appropriate to prefix one's dowsing work with the affirmation "This is for the highest and greatest benefit of all concerned."

Consequently, dowsing is a serious matter, a spiritual matter, and is not to be taken lightly. Obviously when you are first beginning to dowse, you may not know if your interest will wane, or if you will continue to dowse as a hobbyist, or if you will continue as one of service to others. This is fine, and all a part of your self discovery.

As you evolve and become more experienced with your dowsing, you will desire to increase your effectiveness and accuracy. You will strive to become clear and whole; in this way, you will be a channel for the greatest good and benefit. This will be accomplished by committing to advance in this skill, by learning self-mastery techniques, and by gaining dowsing experience through practice (and more practice).

For you to become a successful dowser, you will need to learn to develop your communications with your Intuition. You will need to learn about the basic dowsing tool—the pendulum—and how to invite your Intuition and the pendulum to interact.

The Brain—The Intuition

Dowsing is a way of relating to the earth and the cosmos. It is a way of using the full potential of the brain. We of western society have been conditioned to rely on the rational or left part of the brain almost to the exclusion of the right, or creative/intuitive, side of the brain. What this means for most of us is that our lives are filled with goal-oriented, action-oriented, linear, logical, verbal thinking, at the expense of our imagination, creativity, and receptivity. The right side of the brain, home of the Intuition, has been tremendously undervalued, and in some cases totally ignored. From earliest childhood, most of us were praised and rewarded for performing mental feats involving logic, memory, and other measurable cognitive skills. Our traditional education system is based on the belief that quantitatively measurable skills are superior to skills of the mind that are experienced qualitatively such as imagination and Intuition. The difficulty is that along with this rational system come limiting parameters of reality: Trial-and-error, seeing is believing, cause and effect, and logical, step-by-step reasoning. Unfortunately, for some, the rational mind has been developed so heavily with limiting beliefs that there is no more room in it for any other ideas, and, there is usually no connection with the Intuition. The old approach, "If you can't measure it, see it, hear it, or touch it, it can't possibly exist", is the first belief system to discard for intuitive success. Logical/rational thought has become so dominant in our society, it is hard to believe that some cultures strongly encourage their people to look within for the answers to the unknowns that are beyond the five physical senses.

The Intuition may sometimes seem elusive to you. This is because the Intuition occurs in a part of the brain that has no language. To successfully access the Intuition, you must use all aspects of the brain. You must realize that both your left and right hemispheres of your brain are always working and that you can think rationally *and* use your Intuition at the same time. Again, this is not about one side of the brain being better than the other, it is about a *co-processing*; that you can learn to allow your left brain to put your intuitive insights into words or nerve-muscle actions with the pendulum. Then, allow your feelings to help you to gauge whether your sense of a particular situation is correct or not. Dowsing is the perfect marriage of communications between the two sides of your brain. You have the left side, which loves rational, linear thinking, that wants to know;

what, when, why, and how. It's sort of a sophisticated computer. The intuitive part of the brain does not know how to frame questions, it does not have language skills, simply *feeling* skills. Language skills are the forte of the left brain. In using the Intuition, it is necessary for the left brain to formulate the words of the questioning in a precise and constant form. In dowsing, your answers are only as good as your questions. In other words, precisely communicated questions presented to your Intuition can provide you with the information to which your rational mind may not have access, and your rational mind can then put that information from the Intuition into words or actions.

Establishing communication lines between your rational left brain and intuitive right brain first begins with the desire to Know. Then, allow your rational mind to frame a good question that can be answered by *Yes, No, Maybe,* and various degrees of greater or lesser than *Yes/No.* Allow yourself to observe your intuitive response through the language of your pendulum.

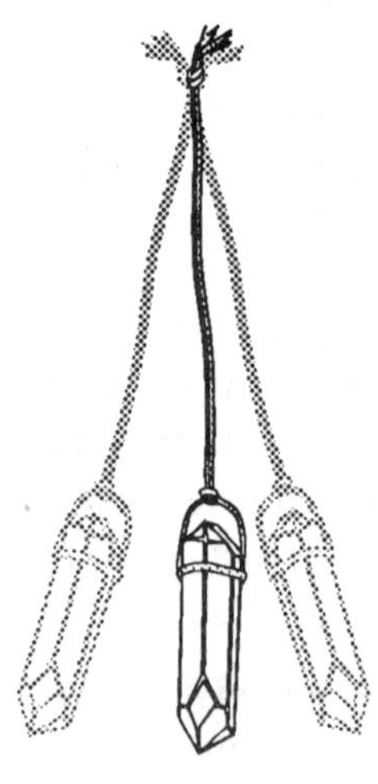

The Pendulum

The pendulum is the one of the oldest, most popular, potentially the most rewarding, and historically the most misunderstood, of all the tools of the Intuition. More than any other instrument, it is the one that is most commonly associated with mysticism, phenomena, and occult. There are some books that talk about the pendulum as a mysterious device that will lead you to untold fame and fortune, personal fulfillment, and romance. Although many of these claims are sensationalized, there may be some truth behind them, although it is distorted by gross exaggerations, appeals to the curious, or flattery of the ego. Of course, this limited approach appeals only to the superficial part of human nature. Because of such claims and distortions, we need to remind ourselves that a pendulum is nothing more than an extension of our Intuition. A pendulum, whether it is crystal, wooden, metallic, or plastic, is simply a balanced weight on the end of a string or chain.

Often we hear people talking about their pendulum as being that which is giving them the answers. Some people personify their pendulum as being that, in itself, which is showing, or 'picking up' information, or having some degree of mysterious independent intelligence. We need to remind ourselves from time to time that it is our mind, the Intuition and the Infinite source (however you may perceive that to be), which is supplying the answers. The pendulum does nothing at all in and of itself. Any response shown by a pendulum comes from you, through you, using your intuitive ability. **You** are the one in charge.

Pendulum Types

A pendulum is a small weight suspended from a chain, or string, preferably about 3-6" in length. The types of pendulums vary immensely, and the various styles can have differing results. A pendulum can be made of almost any substance. Many objects work well for a pendulum, including lead crystals, quartz crystals, wood, plastic, and a wooden plumb bob. Frequently, pendulums are designed with the lower half of the pendulum serving as a kind of pointer.

Some pendulums do have a greater aliveness or respond with greater action, depending on their shape, mass, and size. The important factor is that those that work best have a weight at the end that is centered or balanced so that it can swing freely in all directions without any built-in biased directional influence. The only factor that will make one pendulum work better than another is that it is more comfortable or more appealing to the dowser. Find a pendulum that feels comfortable to you. Remember, its' purpose is to assist you in externalizing your Intuition; it is *not* the source of your answers.

When searching for a particular substance such as gold, silver, oil, etc., a special type of pendulum that has a screw top with a hollowed out inside can be used. This type of pendulum serves as a receptacle for putting a *sample* or *witness* of the substance, object, or thing for which you are searching (see distant dowsing).

Ways of the Pendulum

There are basically four ways in which to use the pendulum:

1. The pendulum is held directly over an object or body and the questions are asked concerning this object or body.

2. The pendulum is held over an object (like food) or over a remedy which is being held in your left hand and questions asked regarding the object's relationship to yourself, or another.

3. A screw top *sample* pendulum is used to dowse for a like substance. The substance is placed inside the pendulum as a sample or *witness*. (an object, substance similar to the object of the search.)

4. The "Witness Method" makes it possible to work on testing objects, substances, or measuring variables for yourself or someone else by means of samples of maps, photos, diagrams and/or *Pendulum Charts*. (See Pendulum Applications and Distant Dowsing for further details).

The Witness and Sample pendulums are often hollowed-out so that the "witness" can be placed inside. Items of *witness* can be water, oil, ore, mineral or a fragment of something approximating the object of one's inquiry. If one is looking for gold, then a gold object or nugget, inserted into the pendulum cavity, will aid in the dowsing quest.

A *'sample'* is a more definite object or substance. It is an item associated with the place, person, object or *thing* sought: such as part of the actual treasure, or a sample ore specimen previously found at the suspected mining site. In the case of a person, the *sample* can be a lock of his/her hair, a bit of fingernail, or an article of jewelry or clothing. When a pet has wandered away, the animal's hair, a collar, or the pet's toy will suffice.

The *witness* or *sample* functions in accordance with the ancient laws of *sympathetic attraction* (like attracts like.) Remote-locating through *sympathetic attraction* relates closely to the phenomenon of *psychometry,* a method used for gathering information about, and from the object being questioned. *Psychometry* relies on the memory thought form held within the object itself. Most people who do psychometry believe that objects still maintain mental, physical, and environmental memory thought forms that can be accessed by our intuitive mind and then relayed to the conscious mind. The psychometrist attempts to determine the object's appropriate frequency or resonance 'in time' through inward, mental visualization. This activity, with eyes closed, is similar to viewing a TV or movie within one's mind. Once the appropriate frequency or resonance is attuned, one often is able to see the whole picture of the persons, places, and events connected with the particular object. Amazing case studies have validated *psychometry* as having scientific value. The relatively new scientific study of psychic archeology is developing with dowsers, psychometrists, and mediums contributing to successful archaeological explorations.

Choosing a Pendulum

Trust your Intuition. The choice of the right pendulum, and the subsequent success in its usage, is in your ability to believe in yourself. Allow your subtle feelings to make the proper choice. We all have these intuitive abilities. It is important that you believe that the pendulum you choose will work for your 'highest and greatest benefit.'

When you choose a pendulum, you may want it to be most appropriate for yourself or for the particular dowsing applications. Consequently, you may choose a pendulum with the added features of being a quartz crystal, having a witness cup, pointer, and precision balance. Some people find that different pendulums for different applications seem most appropriate. Listen to your feelings and then decide for yourself.

Experiment with the string or chain to find the most comfortable length that works best for you. The shorter the string, the faster the action. A heavier weight requires a longer string. Large, heavy pendulums produce less erratic movement while walking. However, you must walk considerably slower when using a large pendulum. This gives the instrument time to overcome inertia, and to change directions; otherwise you may find yourself walking over, or past the energy field emanating from the object of the search.

Chapter 2

Pendulum Language

The pendulum acts as a direct communication device between the conscious and subconscious mind. In this way, it serves to externalize the Intuition. In order for this to take place, we must establish the language between the two parts of the brain.

Pendulums are incredibly versatile in the way that they communicate with your Intuition. Consequently, they can respond in numerous ways. This can cause some unnecessary misunderstanding and confusion. This is because many people will use the pendulum without consciously developing a precise and consistent form of communication, a language between the conscious mind, the subconscious or Intuition, and the pendulum. (Remember that any response shown by the pendulum comes from you, through your intuitive abilities. **You** are the one in charge.)

Developing the pendulum language is a matter of correctly training your conscious mind, Intuition, and body to have many automatic responses that will cause the pendulum to move in a variety of pre-determined patterns so that the shades of meaning beyond the *"Yes"* or *"No"* answers can be determined. Most often, answers fall into the category of being greater or lesser degrees of *"Yes" or "No"*, rather than strict *"Yes "* or *"No."* answers. This makes developing your language with your pendulum very important.

Pendulum Attunement & Language Development

"Identifying" means setting up the language between your conscious mind, sub-conscious mind or Intuition, the energetics of the body, and the subtle muscle movement of the hand and fingers that will cause the pendulum to move. The language is comprised of six directions; forward and backward, side-to-side, clockwise and counter-clockwise, ovals, or a motion in-between forward-and-backward or side-to-side movement.

To begin using the pendulum, first hold the string or chain in the hand you use to write with. (This is your dominant hand.) Hold the string of the pendulum between your thumb and forefinger with your wrist slightly arched so that the pendulum can have a direct drop. Extend and separate the last three fingers of your pendulum hand so that the fingers

do not touch each other. This allows the extended fingers to act as an antennae. Feel if this is comfortable for you. For some people, using the opposite of the dominant hand may be more appropriate. Experiment with both hands until you find which hand is more comfortable and which works more effectively for you. The pendulum can now move easily back-and-forth or side-to-side.

Your first step is to attune yourself to the pendulum. This means that you must find the position on the string or chain that will result in the pendulum having the greatest amount of natural swing for you. Start by holding the string close to the weight at the end of the pendulum. Begin to swing your pendulum back-and-forth as you *slowly* let out the string with your thumb and index finger. After you have let the string out an inch or two, allow the pendulum to circle in either a clockwise or counter-clockwise gyration. Sometimes the pendulum will seemingly go into one of these motions without your assistance.

Pendulum Language Techniques

Now that you have attuned your pendulum to choice of hand, string length, and a sense of it being in motion, it is time to establish the language between your conscious mind, Intuition, and pendulum.

- Let's start by finding a quiet place, free from any distractions. Sit in a chair with your body in good posture, your knees apart.

- All pendulum work begins with an oscillating position referred to by dowsers as the neutral, or "search," mode. This is a back-and-forth swing of the pendulum. Hold your pendulum between your knees and start it swinging gently away from and then directly toward you (back-and-forth).

- In the beginning, if your pendulum doesn't seem to go back and forth of its own accord, make it do so.

The *Yes* or Positive Response:

- Now hold the pendulum directly over your right knee. The pendulum will start to circle in a clockwise direction. This is a *"yes"* response. Clockwise is *Yes.* The clockwise rotation of the *"yes"* response may also be referred to as positive, (+), male, solar, active, or yang.

The *No* or Negative Response:

- Now hold the pendulum directly over the left knee. The pendulum will start to circle in a counter-clockwise direction. This is a *"No"* response. Counter-clockwise direction is *No.* The counter-clockwise rotation of the "no" response may also be referred to as negative, (-), female, lunar, passive, or yin.

The pendulum responds in these ways because we are a composite of all the above aspects, not necessarily one or the other. This is *polarity awareness.*

- If your pendulum doesn't seem to gyrate in a clockwise direction over the right knee and a counterclockwise rotation over the left knee naturally, then make it do so. All we are doing is setting up the language between your conscious and subconscious mind. For a minute or so, hold the pendulum over your right knee, watch it rotate in a clockwise direction, and say to your self, "This means *Yes.*"

- Now hold your pendulum over your left knee. This time, the pendulum will start to rotate in a counter-clockwise direction. Again, if your pendulum doesn't seem to want to gyrate in a counter-clockwise direction, make it do so. Continue holding the pendulum over your left knee, watching it rotate in a counter-clockwise direction and say, "This means *No.*"

If your are already using a different language than the one outlined above, that's fine. Continue to use the one you have already established. It doesn't make any difference what language or code you use for your responses, just as long as you are precise and consistent.

Pay no attention to any other movements at this time; program yourself only for a *"yes"* or *"no"* response. The Intuition communicates principally by impressions, imagery, symbols, and feeling when sending or receiving data. Therefore, it is most helpful to instruct your subconscious of the *intent.* The key here is to remember that words don't generally impress the subconscious mind; you must send the *"Yes"* with every available impression and feeling possible. For example; use body language with a nod of your head with *"Yes"*, and side to side for *"No".*

The next step in developing your pendulum language is to apply your established directions and to expand these parameters to a linear perspective. This may sound more difficult than it really is. You can determine all six pendulum language directions with the following exercise:

- Draw a six inch circle on a piece of paper with a crisscross in the center, dividing the circle into four sections. (See "Pendulum Charts", and use the *Pendulum Language Chart*.)

- Give the pendulum a 3-6 inch length on the string and hold it directly over the intersection on your circle or the *Pendulum Language Chart*.

- Look at the chart, remove your attention from the pendulum; focus the intention with your mind to establish the pendulum language.

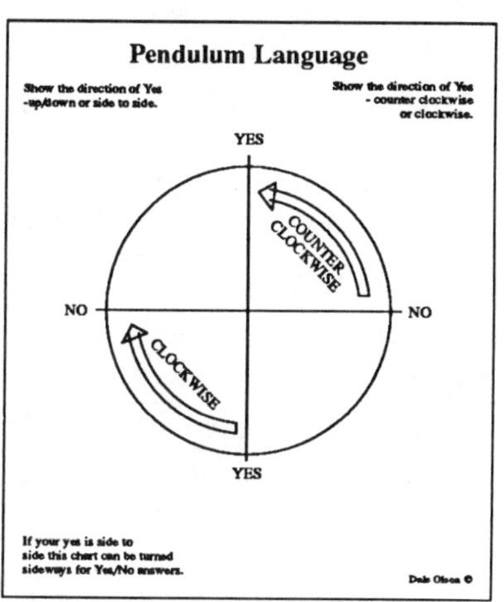

Knowing Your Intuitive Mind: Pendulum Charts
Dale Olson, Crystalline Publishing, Eugene, Or.

- Being precise, specific, positive, and affirmative with the communication to your subconscious mind is absolutely imperative for success and accurate results. It helps to speak the intentions out loud. For example, "Intuitive Mind, please indicate the direction of *'Yes'* for me." Again, for most people this would be indicated by either a backward-and forward swing, or a right-handed or clockwise gyration. For most people a side-to-side swing, or a counter-clockwise direction indicates a *"No"* answer.

- A swing that is in between can indicate more-or-less, *"maybe"*, or *"rephrase the question"*. Direct your mind and pendulum into action by verbally making your statement; "Intuition, or

Infinite Intelligence, indicate to me which direction is for *"Maybe,"* *"more-or-less",* *"rephrase the question"* and so on. Most often this is indicated by a swing between the back-and-forth and the side-to-side direction (a 45 degree angle to the crisscross on your circle).

Within a short time you will reach a stage where you can look at the pendulum and ask it to indicate *Yes, No , Maybe, or More-or-Less* and the pendulum will respond immediately.

If at any point there is little or no movement, relax, breathe in and out several times and try again. Sometimes patience is needed for the communication to become established, and, for some people the pendulum responds immediately. If you should get slow answers, it can be helpful to cup your hands around running water, or place a quartz crystal in your hand for few minutes. This will help you to increase your energetic level to facilitate more success with your pendulum work.

If you state a question and the pendulum does not move at all, or it moves inconsistently, then consider:

- are you tired?

- forcing the situation?

- focusing too heavily on the pendulum itself?

- not focuing on the objective?

- do you feel any mental or emotional imbalance?

- is there a more appropriate time?

- is there a more appropriate place?

- is there a more appropriate way to state the question?

If any of these points ring true, then give it a rest and approach it at another time.

If there is a problem with making it move, perhaps it is a matter of too much focus on the pendulum. In other words, there is too much attention of the mental and/or emotional mind involved in trying to make this process work. Most often, success requires a defocusing of the mental and emotional mind. This defocusing can sometimes be achieved by holding your elbow tucked against your waist to the side of your body, wrist arched with pendulum hanging freely so that the pendulum is seen more out of your peripheral vision than by direct sight.

Erratic patterns are difficult to figure out, however, the confusion is not with the instrument, it is within your own mind and the clarity of how you stated your questions. Your answers will be only as good as your questions. This is why we recommend that you keep your pendulum language simple at the beginning and remain focused exclusively on the object of your questioning.

If you have experienced difficulty with the use of the pendulum, don't give up. Anyone can dowse if they are open to it. Keep on working on it. Do these first two exercises, (establishing movement over and between your knees, and the Pendulum Language Chart exercise) for seven minutes every day for a week. You will excel rapidly if you practice this approach until the movement becomes second nature. All that is required is a relaxed state of mind and practice, along with the willingness to act as though the pendulum were responding independently. If nothing happens, go back to moving it consciously and reminding yourself which motions means *"yes"* and which ones mean *"no."* If tension or frustration begin to take over, stop for a while and resume again later.

If you still have problems, look at your belief systems, learn how to balance and defocus the mental mind, choose a different location or time to do your pendulum exercises, and make sure your area is free of distraction. (See mental and emotional interference.)

In addition to the *Yes, No, Maybe, or More-or-Less* query, the pendulum language may be expanded by means of numbers, alphabet letters, and positive (+) or negative (-) response. Other refinements will develop as your communication skills improve.

Number Priority

This form of pendulum language establishes a way to use numbers as a means to measure or gauge any variable for purpose of comparison between two subjects. This can apply to anything from consumer products to services, events, or any other form of choice-making.

Begin by carefully establishing in your mind the difference between the *Yes, No and Maybe* replies. Then, when you're ready, the following method will assist you to find the correct number answer(s).

- First create a question in your mind about something simple that you would like to measure. For example: "Intuitive mind, or Infinite Intelligence, the food in front of me is to what degree_______ for my highest and greatest benefit?"

- Decide beforehand the number scale you wish to work with. For example: you can count it on a 1-10 scale; or multiples of ten (10-100, or 100-1000). Keep your mind clearly focussed on the particular number scale you are counting AND the object of your measurement.

- Start your pendulum oscillating in the *Maybe* position between the forward-and backward and the side-to-side direction. Present your question to your Intuition. For example, using multiples of 10, start your count 10, 20, 30, 40, 50, <u>60,</u> and the pendulum begins to deviate from the *Maybe* position moving towards the *Yes* mode. Now start counting backwards until the pendulum resumes the *Maybe* direction... 60, 50, keep going to 40 and, again the pendulum begins to deviate going back to the *Maybe* mode. Now go back to <u>50,</u> and sure enough the pendulum resumes the *Yes* position. In this case, '50' is the correct measurement or answer to the question. This will be discussed in greater detail in the sections on how to set up your parameters of questioning and working with the *Percentage Charts*.

Alphabet Mode

Developing your pendulum language to also include the alphabet will assist you to determine unknowns through finding a letter, or spelling out a name, word, phrase, etc...

- Start the swing of your pendulum in the *Maybe* position (between the forward-and-backward and side-to-side positions.)

- Mentally create the question: "The name of my guide or teacher begins with the letter A...? B...? C...? etc....

- Watch for any change in the pendulum's motion. For example, A...? B...? C...? D...? E...? F...? G...? and the pendulum begins to deviate from the *Maybe* mode to the forward-and-backward position...keep going to H...? and again, the pendulum begins to deviate to the *Maybe* position again. Now go back to the 'G', and sure enough the pendulum resumes the *Yes* position. In this case, <u>G</u> is the correct first letter of the name or the answer to the question. If you have developed an alert dowsing sensitivity, your pendulum will respond readily to your questioning.

With patience, you can develop letters into names, words, phrases, addresses and even sentences. In this way you can form word meanings from the subconscious mind through the letter-symbol language. For additional information see the section on *Alphabet-Numerical Pendulum Chart*.

Checking Objects Pendulum Language

As you become more proficient with the pendulum and the language between your conscious and subconscious mind, you will find a system that works for you. Whenever you hold the pendulum over any object, animate or inanimate, you will pick up the radiations from it and the pendulum will usually begin to oscillate.

- Begin with the *search* mode of a forward-and backward swing. This signifies the neutral swing, which merely puts you in contact with the object.

- After you ask a question in relationship to the object, the pendulum will start rotating. For most people, if it rotates to the right or clockwise, it is saying *"Yes"*; this is the positive swing, indicating a positive or harmonious condition. If it rotates to the left, counterclockwise, it is usually saying *"No"*; which is a negative swing, indicating a negative or inharmonious condition. The degree to which the pendulum rotates will indicate the extent of your answer. If you ask a question which does not have a yes or no answer, or, if the question is confusing, or if it doesn't have an answer at all, then the pendulum will oscillate side-to-side indicating a *"No"* answer. You may have made a mistake in how you asked the question. Try rephrasing your question.

Now that you have established your pendulum language and directions, let's do some simple testing and practice exercises.

The first exercise is for direction and polarity check:

- Hold your pendulum over the top of your left hand. If working properly, the top of your left hand will cause the pendulum to gyrate in a clockwise motion indicating a positive energy.

- Now turn your left hand over, palm up and check; this most likely will cause the pendulum to gyrate in a counter-clockwise motion, indicating a negative receptive energy.

You will experience the opposite findings if you hold the pendulum in your left hand and use your right hand for the test.

This second exercise is for accuracy:

- Create five separate papers with mathematical calculations, some correct and some with errors.

- Mix them up, place them face down on a table, and find the correct ones through positive (+) or clockwise gyration of your pendulum and the incorrect ones by negative (-) or counter-clockwise direction.

This can be a helpful test. However, we find that our abilities work best when we have a good reason for working with the object being tested. The Intuition itself does not like to be tested. If you are trying to trick the Intuition or question its validity, then you will most likely get inaccurate answers. This is an indication of lack of trust and your answers will reflect this. These tests are a means of developing accurate communication lines, and need to be viewed as exercises for developmental purposes.

The third exercise is a self-tuning or sensitivity test:

- This is a good test to do at various times under various conditions. Stand erect, facing due West, relax as much as possible, place your left hand directly over your solar plexus (area from bottom of rib cage to a few inches above navel), palm inward with the fingers closed.

- Suspend the pendulum from the right hand, using the full length of the string, so that the pendulum is opposite the center of the left hand and about 8 inches out from the left hand.

- Due to the polarity in the area of the solar plexus, the pendulum will start to gyrate in a clockwise direction. Count the number of gyrations carefully. When the pendulum's movement begins to show a noticeable change in its direction or behavior, make a note of the number of gyrations prior to the pendulums change.

The degree of sensitivity and strength is indicated by the number of gyrations, and although they vary considerably from individual to individual, they can be grouped roughly as follows: 15-30 weak, 30-50 medium, and 50-100 good. These figures represent the number of complete gyrations and not per minute. If the gyrations are weak and less than fifteen, your chances of getting totally reliable results are remote. Try another time or place.

This is an interesting test, it indicates that the human body is extremely sensitive to outside influences, and can be measured accordingly. With practice and patience, you will notice an increase in your dowsing abilities and sensitivity.

This fourth exercise is for finding direction:

When you are outdoors, your pendulum can be used to indicate to you where North is.

- Hold your left arm up, and if you have a set of keys, hold them in your left hand between your first and third finger, dangling downwards (which will act as an antenna), and point with your left index finger.

- Hold your pendulum in your right hand, and slowly turn around in a circle. Start the pendulum swing in the *Maybe* mode position. Pay close attention for any deviation of swing with the pendulum as you turn your body The pendulum will start to move to the *Yes* position as you approach North. If you go past the North position, the pendulum will again start to deviate back to the *Maybe* position.

The pendulum indicates direction because you are working with the electromagnetic energy of the earth, which runs North and South. Do check this technique for accuracy with a compass.

Once you have established the communications, or pendulum language, with your Intuition, you can expand your development by working with questions to which you already know the answers. With practice, you will acquire a feel for the pendulum, and we suggest that you do not go on to more advanced questions until you consistently get accurate results to questions about which you already know the answers. In this way you, will establish a consistent language with your Intuition. It takes time and patience to develop these abilities because it takes time for the subconscious mind to tune into the information requested and for the different parts of your being to become familiar with the language of the pendulum and the various aspects of your consciousness.

You may find it helpful to keep a logbook to record the results of your exercises, experiments, or experiences. You may record the time of day, the direction you were facing, the specific questions you asked, and your pendulum answers. This way, you may refer back to the questions to enhance your communications or to discover errors. The record can also help you to discover the areas in which you are most accurate or proficient. The logbook may also reveal a great deal about your continued study and practice.

When first learning how to work with a pendulum, you may find yourself giving more attention to the movements of the pendulum than with other tools of the Intuition. This is fine, because it can act as a biofeedback device for identifying that focus and defocus state during intuitive insight. With this type of state identification, you can attune yourself to very subtle and abstract levels of Intuitive insight not solely based on external factors. It is especially important to be very clear with your intent and the questions you ask when you use the pendulum, and to release or neutralize any concepts or assumptions that may reach or limit your conscious or subconscious mind.

Chapter 3

Successful Form

How Dowsing Works

There is no magic about dowsing. Most, if not all, of us have this faculty. The dowser, with the aid of a physical instrument and a previously determined language, poses a question to his/her subconscious mind for the purpose of gathering unknown information — at a distance— regarding a person, plant, animal, substance, time or place. The answer comes from the Intuition, beyond the limitations of one's logical and reasoning processes, though one's intellect does initiate the activity and eventually summarizes the findings.

From one point of view, dowsing involves neuro-muscular action. Automatism, as applied to dowsing, implies that thought energy registering in the brain travels through the nervous system. It also suggest that we have the ability to receive rays of light or waves of sound that surround all of us, and all objects. These impulses are then sent down the ulnar nerve causing muscular reflexes in one's fingertips to move the pendulum beyond conscious physical control. This concept indicates that the pendulum is only a tool: a symbolic physical device moved by the forces of focused, mental and neuro-muscular energy. Scientific tests have shown that there is an increased electrical tension on the skin of a dowser when he/she approaches the object being searched for.   Again, the subconscious mind is able to locate the object and send its message to the conscious mind by creating the electrical tension that causes the muscular movement which makes the pendulum respond.

Another theory on how dowsing works is that of *sympathetic resonance*. This theory suggests that traces of gold (for example) within the dowser's body will resonate with the gold in the ground. This specific *"resonance"* presumably can be felt down to the cellular level. Consequently, a sympathetic vibration and energy flow between the two sources of gold and the subconscious mind create a neuro-muscular response that activates the pendulum. The range of this technique appears limitless. This theory implies that as long as the dowser's inquiring mind makes the connecting link between like substances, the distance is unimportant.

The human body is electrical and magnetic in nature, and consequently, it is sensitive to and is able to register the multitude of electro-magnetic forms of energy. When you think about of being able to attune yourself or to come into sympathetic resonance with something, it helps to trust this extraordinary capacity that is already a part of your being.

Communication Skills—Forming Questions

Movement and manifestation is dependent on communication. We are in continual communications with ourselves, life, and all things outside of ourselves. One can communicate with any life-form, object, or substance providing that the visualized thought form(s) are held in one's mind-focus.

The secret to successful dowsing comes in one's ability to clearly hold the mentally posed question in your mind. The clarity of the query is foremost: Incomplete, ambiguous, or hazy questions will bring about confused answers. The inquiring mind should frequently re-state the question, preferably one inquiry at a time. Multiple questions by the unskilled pendulist frequently produces conflicting information.

The main keys to your success will be keeping yourself objective and detached from the outcome, and being extremely precise with your questions. In your communications, your answers will be only as good as the questions. It does require certain knowledge on the subject in order to know what questions to ask. Your questioning will work best when the questions are stated in a positive declarative format, such as:

Now is the best time to do this reading?

This is the best place to do this?

My intuitive accuracy is what % today?

It is to my highest and greatest benefit to ___________ at this time, at what % _____?

This plant is in need of _____?

The lost object is what direction or degree from this point?

This body is in need of what remedy for this situation?

Asking the *right* question in the *right* way is 90% of the dowsing skill. (For further details regarding questioning, see the *Pendulum Chart* section.)

Of course, timing, place, and emotional-mental balance are equally important variables. Always ask, "This is the right time and/or the right place to be asking this question?," or, "This is the appropriate question?" If NO, then wait for another time or place, or go on to another question. By asking the questions, "This is the right time, place, or appropriate questions?", you establish a trust and respect with your Intuition. You also must respect others' private space.

We need to always be certain that we have permission to dowse from what we perceive as the Source of the Highest Good. This is to make sure that we do not disrupt the harmony of the space in question. Remember that whenever we do dowsing that involves someone else, we will be entering that person's private space. Whether it be another individual, their property or land, it is imperative that we do not invade these private spaces, and so we follow a sequence of permission. We dowse for that permission with a threefold question: ***May I, Can I, Should I.***

- ***May I*** seek this information—that is, am I permitted? Good intention are not always enough.

- ***Can I*** receive a clear answer that will allow me to be of appropriate service—am I able, do I have the skill? Being good in one form of dowsing does not assure accurate results in another.

- ***Should I*** seek it— is it the appropriate time and am I the one to do it? It may be that a particular type of dowsing is inappropriate for you to do at that particular time. Even with right intention, one can still do harm without realizing it.

We need to realize that there are some circumstances or situations which are intended to be as they are for reasons beyond our awareness or understanding. What may appear as a nasty situation may prove to be a miracle in disguise later on. By dowsing for permission, we are asking for the awareness beyond our reasoning. Though we may not always understand why, we must learn to trust the permissions.

Your approach to questioning with a pendulum requires that your mind be in a neutral, quiet state, never letting thought or desire interfere with, or influence, the answers you seek. It is important to keep an "I don't know" attitude. If a condition was negative yesterday, I really don't know about it today; it may be the same, it may have improved or regressed. You must be aware of what your mind is doing at the time of the questioning. Any thoughts about a possible answer, any personal desire, and any ego involvement or tendency to show off will influence your work. However, if you are centered and balanced, and your questions are completely objective, you can trust your intuitive answers.

After you have worked with the pendulum for a while, you will gain confidence. Start with some simple exercises, and move on to more complex work. When you finally proceed to areas where you may be emotionally involved, you will find that you can trust your objectivity with them to a greater degree. However, when doing personal questioning and you have an investment in the outcome, we strongly suggest that you see the answers as, "Oh, this is very interesting, but, we'll see." If you see the same answers coming up over and over, month after month, then there may be validity to the acquired answers.

With knowledge comes power, and to have access to the Infinite Knowledge, Infinite Intelligence, Infinite Wisdom greatly increases your responsibility. It will also test your ability to avoid the traps of the ego (for example, becoming enraptured by this sense of personal power.) Such ego entrapments can be avoided by prefixing all of your pendulum work with a starting affirmation that asserts that you are dowsing for the greatest good for yourself and others. Then follow with the threefold permission.

- "Infinite, or Infinite Spirit, allow this work to be of the very highest and greatest benefit to me."

- "Infinite, or Infinite Spirit, allow this work to be for this person's ________(name)________ highest and greatest benefit, their need is ____________________?"

- "Let this be of thy will, not my will."

- "I invoke the Light within. I AM a clear and perfect channel. Love and Light are my guides."

- Then ask the threefold permission:

"May I?"— Am I permitted?

"Can I?" — Am I able? Do I have the skill to do this?

"Should I?" — Is this the time? Am I the one to do it?

(You may also add, "Is this the truth?" as a final check.)

If you do not get permission for all three, DO NOT continue on. You may try again at another time or place, or when you are in a different mental-emotional state of mind.

Subtle Sensing

As we have discussed, it is important to establish precise and consistent language communications with the Intuition. After some development of the dowsing skill, sensing the answers to your questions can often be experienced in more ways than movement of the pendulum. Seasoned dowsers do not just see the answers in the movements of the pendulum, they also feel or sense the answers in terms of frequency registrations in their hand, arm, or entire body. Some extremely physical dowsers can feel subconscious contact and the presence of information in answer to their question(s) through a sudden subtle sensation of an itch, tingle, change in body temperature, ringing in the ear, feeling of nausea (pressure, discomfort) associated with solar plexus responses and, in some cases, dizziness. These types of sensations can be incredibly accurate in letting you know through physical response the answer to your line of questioning.

The subconscious understands the request through feelings and impressions from the intuitive mind, and sends its reply back through the dowser's sympathetic and parasympathetic nervous system. The dowser may be attuned to not only the pendulum movements independent of conscious physical control, but also he/she may be aware of subtle-sensing that is experienced in other parts of the body. This, like your pendulum language, can also be an effective form of communication. For further development of subtle-sensing and applications of it's use, see <u>KNOWING Your Intuitive Mind</u> by Dale Olson.

New Communications

In your pendulum work, you may find sooner or later that your existing pendulum language may be inadequate in handling a particular problem or application. You may find yourself needing to improvise by using convention. Convention in dowsing implies an agreement between one's conscious and subconscious mind. It may be a new experience, subject to change, and an improved procedure meeting the often unexpected dowsing need of the moment. It is a matter of responding to spontaneous inclination to use a new pendulum pattern that may access additional information. For example, you may be locating a situation where you know you are near the object of your quest, and you suddenly find your pendulum behaving in a new and unexpected way: attempting to make a figure 8 or rotating in a right or left oval. What does it mean? Is the subconscious mind trying to tell you something beyond the range of your present pendulum language? Take the time to meditate upon its' meaning. See if you can sense an impression or visualize an image on your mental screen. You can develop new additions to pendulum language by working through and expanding on programs of this nature. These new pendulum movements are the result of convention, and are additional to one's established pendulum language. Once your pendulum language is established, it should never be altered. Once it has been ingrained into the subconscious mind, you may expand upon it, but not change it. For thousands of years the master teachers have uncovered the foolishness of changing positive symbolic meanings that have been impressed upon the unconscious or subconscious matrix of one's mind.

Mind Preparation

Approach the pendulum with *enthusiasm and with confidence* that you are able to do it, for a half-hearted or doubtful attitude will only result in uncertain findings. Always work in quiet surroundings, by yourself if at all possible, away from skeptics, negative thoughts, or anyone trying to influence you. See yourself, the operator, as a super-sensitive receiver. This will help in your development.

Because we live in less then a positive, harmonious, and secure world, we feel that all work of this nature needs to be surrounded by the *White Light & Love* shielding. The *White Light* is the Source, the God lifeforce that dwells in all things. It ensures that all is for one's highest and greatest benefit. To make this protective shielding, visualize yourself enveloped by the *White Light & Love*. Take a *White Light & Love* shower. Visualize yourself drawing in the *White Light & Love* through the bottom of your feet; letting it go up your legs, up your spine, up and out through the top of your head, and then let it shower down all around you. Maintain this glow of protective *White Light & Love*, and feel the unconditional love throughout your being. This universal method of shielding oneself from psychic intrusion has been used throughout the ages to protect oneself and others from outside influences. This technique, along with some of the starting affirmations and the threefold permission, will greatly assist you in keeping a protective field around you and others you may be working with. It also balances the mental-emotional mind.

The secret of mind-power and successful dowsing arise from concentration, and one's ability to hold a particular thought-symbol or visualized idea continually in focus; fixing it in the mind, and then releasing the idea, or image to the powerful realms of one's subconscious or intuitive levels of the mind.

Self-conditioning and a positive frame of mind are essential for successful dowsing and other intuitive work. Before you go to work on a project, we suggest that you spend several minutes doing the White Light & Love shower and self-conditioning. For example, if I was dowsing for gold, I would go through all of the steps within my imagination. I would visualize the dowsing process step by step through to the actual digging and the finding of the gold. Practice makes perfect. I work on self-conditioning everyday. I mentally go over each project or task in my mind. It is easier for most to visualize a step by step project than to see some of the more intangible tasks that we often present to the Intuition. In applying this technique to the construction business, I saw that for every hour I visualized in this way I saved three or more hours of work on the job site with less error and therefore greater success.

To apply this technique one step further in locating the gold example, I take the time to prepare my mind before going out on a site. I visualize a mental tour of the land to which I'm going. I have found that with this technique, I am not only creating a connection to the location within my imagination, but I am also attuning my perception to what the physical terrain looks like, and feel the energy resonance of the area and all things within it. In other words, I use all of my senses to connect with the area in question. While on this imagined tour of the land, I first establish an arbitrary starting point (tree, rock formation) in relationship to the lay of the land. I then look around for the minerals. I pretend I am looking into the earth and objectively searching for the gold bearing ore. When I spot the ore, I make a mental note of its location in relationship to the earth's surface. I then complete my mental experience by visualizing the ore being successfully mined and brought to the earth's surface. Most often, when I visit the site, the land will look exactly as I had visualized, and the ore is usually located in the spot that I had mentally envisioned. This basic technique can be used in just about any application you can imagine. Another example would be in finding a water well location. Mentally standing at the front door of the house, I would ask: In what direction is the best place for a well; how deep is the water; how far from the front door is it; what is the flow and direction, what is the quality of the water and so forth.

Imagination and visualization are extremely powerful tools, and among the most important disciplines taught in all shamanic training. To successfully visualize, the mind needs to be relaxed, balanced, and quiet. Shamans often used peyote, jimson weed and other hallucinogenic plants to induce flights of imagination and to open the way to the inner self. Controlled breathing can produce similar results, in a more controlled and beneficial fashion (see the section on Mastery of the Breath).

Directing and developing your guided imagery will not only assist you in successful dowsing applications, but it will also help you to explore the limitless inner resources of your own consciousness.

All of these pendulum functions point to scientific and measurable methods of accessing the Intuition. But in order to achieve these higher levels, you must get through the basics of form and the clearing of ego responses. The next chapter will introduce these aspects and attributes.

Chapter 4

Emotional and Mental Interference

There are basics in any form of art which must be learned and practiced, and dowsing is more art than science. What happens when an individual doesn't learn dowsing basics? First of all, we have seen individuals who learn some dowsing skills and if they do not immediately get accurate results, they quit using the pendulum and tell their friends how it doesn't work. Second, we have seen dowsing receive criticism as a result of people who, after having been shown how to dowse, pick up a pendulum or dowsing rods and find they can indeed get a dowsing response. Immediately claiming to be experts, they go out and tell their friends to dig here and drill there.....and chaos results. These are examples of people who did not first learn and practice the basics.

By learning and practicing the basics, it is possible to become a successful dowser. But a stumbling block in the development of dowsing skills, and in the development of the Intuition itself, is mental or emotional interference. It is the one thing that eats at the very core of one's trust. Without trust you have nothing. Mental and emotional interference, and additional stumbling blocks which may prevent successful intuitive dowsing, include:

- trivial use

- fortune-telling, predicting the future

- impatience, fatigue

- mental distractions, overconfidence

- undue influence from skeptics

- lack of concentration

- needless superstition or limiting beliefs

- poorly defined communications or language between the conscious and subconscious

- imperfect coordination between one's mind and the instrument.

- imbalanced emotional state of mind: fears, doubts, anger, resentments, sadness, grief, anxiety, greed, jealousy…

Certainly, the pendulum is a remarkable tool that allows us to reach out for answers that are beyond our ordinary intellectual scope. Because the intuitive ability is so extraordinary, it is best not to use it for trivial or minor matters. For example; "Should I have the green salad or the spinach salad for lunch?", or, "should I use a blue pen or black pen to write this letter?" These may seem farfetched examples, but I think we all know people who do ask similar petty questions. Worst of all, we sometimes may find ourselves asking the pendulum for answers for equally unimportant matters. If you use the same clearing technique with your mental mind for more common daily decision-making, (for example, going into a slightly defocused or altered state of consciousness,) you will always access information which is for your highest and greatest benefit. Using these skills more often in your daily decision-making will assist you to make better choices, and it will also develop your intuitive proficiency.

One area of dowsing that is very tempting is predicting the future. Until you have many years of successful, accurate dowsing, it is best to avoid the area of predicting the future altogether.

If you misuse your intuitive abilities in the way that many people have misused the Ouija board and other forms of divination, you will be using the pendulum at the level of a parlour game. If you are trying to find out when you are going to die, when you will find the person of your dreams, or when you will win the lottery, and other types of fortune-telling, it safe to say that you may find yourself greatly disappointed. If you merely play with the pendulum as if it were a game, the results will be superficial and you will only get what is in your own subconscious and mental mind, and not information from your Intuition or the collective consciousness.

When used properly, the Intuition tunes into the universal consciousness where the Infinite Intelligence will bring forth the appropriate information. Those who work at predicting the future know that the time element is the most difficult to forecast. The only time the pendulum, or subconscious, can tell you something in the immediate future is when the information is already present in the human consciousness. You cannot get an accurate reading before people have made up their minds about an event or outcome. When you work objectively, new information and facts can accurately be discovered that were previously unknown to your conscious mind. The information, in order for it to be available to you, must already exist somewhere in your subconscious mind.

Sometimes individuals using the pendulum complain of unexpected confusion, lack of consistent agreement with oneself, alternating *Yes* and *No* answers, erratic responses of the instrument, and outright failure. This is sometimes followed by temporary or partial loss of one's dowsing ability. This type of problem mostly occurs from impatience, fatigue, mental distractions, emotional upset, overconfidence, or undue influence from skeptics.

Superstition can limit your dowsing ability. Believing that magical formulas, including dowsing at a particular time of day, a certain place, only under certain weather conditions, or while you are wearing certain types of clothes, etc, are the *only* way your intuitive abilities will function means you have sold yourself a program. Once you have set such a limitation in dowsing, you may be stuck with it, for it can be very difficult to change. As always, it is best to be mentally and emotionally centered and alert to any belief systems or dogma that may be limiting to your dowsing abilities. Always approach dowsing with an open mind and an objective attitude.

Some people believe that dowsing is a gift only bestowed on them and a precious few. These people are on an 'ego trip.' This belief is an outright fallacy and simply another limitation. Some people do learn and adapt more quickly when adopting these skills. Most skilled dowsers have withstood the tests of time and patience, and most admit that dowsing finesse is usually acquired through some tribulations of trial-and-error and continued practice.

Successful Dowsing Form

- Communications: The right questions, at the right time, at the right place.

- Balance: Mental and emotional balance, a relaxed state of mind.

- Breathing techniques

- Meditation, Prayer, Affirmations

- Practice, Patience, Persistence.

- TRUST in yourself and the source of your Intuition.

Once again, we emphasize the necessity of saying *affirmations*, practicing the *White Light* shower, and asking for the *threefold permission* before doing your pendulum dowsing. Remember that it is good practice to preface your intuitive work with a starting affirmation or prayer such as "Open the way for that which is of the highest and greatest benefit," or "Let this be thy will and not my will," or, "I invoke the Light within. I Am a clear and perfect channel. Love and Light is my guide." Find an affirmation or prayer that works for you and your belief system. Affirmations, prayer, the white light shower, and the threefold permission are extremely powerful in offsetting the interference of the ego, the intellect, and outside influence.

Dowsing embraces the implicit belief that we do what we do through love—for the greatest good of all concerned. This means dowsers must realize what to do with the answers we receive through our Intuition. Not all knowledge that we discover through dowsing is appropriate to pass on. We must look to our Intuition to determine what is appropriate to share.

Chapter 5

Self Mastery

I Now Breathe In Life Fully.
I Relax and Trust The Process of Life.

Mastery of the Breath

Practicing affirmations, prayer, threefold permission, and the white light shower, will go a long way to ensuring that your dowsing is clear, objective, and for the greatest benefit of all concerned. Mastering your breathing skills will assist you to be emotionally and mentally balanced. This is important because effective and accurate work with the Intuition is greatly increased when the dowser comes from a place of emotional balance. Practicing certain breathing techniques can clear negativity, and promote emotional balance and general well-being.

By emotional balance we mean a state of emotional harmony, neutral feelings neither too high nor too low, but centered. This means your thoughts need to be positive without attachment to the outcome, and without any critical or judgmental thoughts toward yourself or others. Remember, it is the ego and negative emotions that are the major deterrents to your intuitive abilities. Fears, doubts, anger, resentment, greed, anxiety, jealousy, etc... can be the greatest hindrances to the development of the Intuition (as they can be hindrances in all aspects of life.)

Of course, we already know that it is imperative to be free of such negative emotions if we want high quality living. Packing these feelings around is not what we would call a good time. But most of us have never been taught very effective ways of neutralizing or transforming this negative 'stuff'. We can release these negative mental and emotions thoughts and feelings with the following breath techniques and concentrated effort. Many of us have had years, or even a lifetime, of poor breathing habits. These habits are largely responsible for most of our imbalanced or negative states. Have you ever noticed, when someone becomes fearful they almost stop breathing and basically 'shut off'. In my therapy practice, I have seen hundreds of individuals who will shut down their breathing when feeling 'not good enough', or 'fearing failure'. This dynamic is quite destructive because it creates a vicious circle of poor breathing habits and patterns of negativity.

Breathing mastery is one of the most beneficial skills that one can acquire. The author feels this section is the most important part of this entire book; it is a way to heal all aspects of one's life. Much credit goes to Dr. Marcel Vogel, Paramahansa Yogananda, Yogi Ramacharaka, and many other authors, researchers and teachers who know and teach that the breath is the key to mental and emotional balance, optimal health, and well-being. The techniques which follow are extremely dynamic. Many of us will need to practice and experience them first-hand to realize the incredible power and healing potential available to us.

In all of the breathing techniques we will share, there are several parts on which to focus. First of all, you will want to create a positive mental image in your mind, your *intention*. *Intend* for your work to encompass the principle "for your highest and greatest benefit". Secondly, relax into the breathing technique which you are practicing. (Learn it well so that you can call upon it at any time). Lastly, let your ego go; trust your Intuition.

This will take some practice, so be patient, willing, and enthusiastic. Remember, it *is* a learning process.

In summary, mastery of the breath is the key:

- to heal all aspects of the physical, mental, and emotional self.

- to balance the emotional-mental mind.

- to oxygenate the cells of the body.

- to draw in Prana (universal lifeforce).

- to release the power of thought.

- to release stress, promote relaxation.

- to neutralize negative and critical thought forms.

- to release old imprints and assist in restoring one's personal power.

- to release fears, doubts, anger, resentment, greed, grief, jealousy…

- to open intuitive channels.

- to create protective psychic shields.

When we breath deeply, we invigorate the entire body through oxygenation of the cells, and we bring in the vital charge of energy called Prana, which is the universal lifeforce. This vital force sends its currents through all body systems and is absorbed down to the individual cells of the body. Your degree of health and vitality is determined by your ability to absorb and circulate Prana. When you inhale deeply, you pull more Prana into your body. Prana is utilized by the mind to build the patterns of thought, being, and intention of what we wish to do, be, or perform. Our intention comes on the indwelling breath. Inhalation draws an electrical charge into the body (negative electrical charge); exhalation discharges an opposite polarity (positive electrical charge). Inhalation draws Prana (life force) in, and exhalation releases the power of thought. We express our thoughts with each exhalation. (If you don't believe this, try carrying on a conversation while only inhaling.)

First Breath Technique for Emotional and Mental Balance

The first technique is for bringing about balance of the emotional and mental mind. This is accomplished by utilizing the dynamics of the breathing mechanism itself. This technique can be used for releasing accumulated negativity, relaxing, and mental clarity.

To use the breath in this first release method:

- Create the intention in your mind to be in a clear and positive state; in a harmonious state of mind. Free yourself from any negative or critical thoughts.

- Breathe in through your nose with this intention and create an image in your mind of yourself in this balanced state.

- Breathe out through your mouth slowly and easily, releasing any and all negative thoughts or patterns. It is best not to forcefully exhale, but ever so slowly release these negative and critical thoughts with blessings, rather than with anger.

- Continue to breathe in through your nose, imprinting positive patterns of thoughts or affirmations toward yourself or others. Breath out through your mouth any negative imprints, patterns, or thoughts. Do this release and balancing exercise, breathing in and out in a connected loop; breathe deeply, evenly, and completely.

This sort of breathing technique can take anywhere from five minutes to half-an-hour to bring the body and mind to a harmonious state of clarity and balance. Again, release negative patterns of thought by breathing out through the mouth. Imprint positive patterns of thought by the breathing in through the nose with intention and positive thoughts and affirmations.

Second Breath Technique for Balancing and Meditiation

To use the second technique:

- Start by sitting with your back straight (you can sit in a chair or on the floor—it makes no difference.)

- Begin to breathe in and out through your nose slowly, deeply, connecting the breath.

- While you're doing this circular breathing, connecting the inhalation and the exhalation, visualize the energy in your spine going up and down like a thermometer with red mercury. With each breath in, the red mercury goes up the thermometer. With each breath out, the red mercury goes down the thermometer.

Do this connected breath only 14 times, as suggested by Paramahansa Yogananda. This technique works best to do twice a day (in the morning and evening) to assist you in maintaining a positive, meditative state. Of course, this exercise can be done at any time to bring you into mental or emotional balance, to initiate meditation, or to create a state conducive for working with your Intuition.

Micro & Pulsed Breath

The next breath techniques have been labeled the Micro and Pulsed Breath. They came into form through Marcel Vogel in his research with adept Yogi masters from India. The material that is to follow is a shortened version; it will take only a few hours of practice to start feeling the empowering affects. Some masters have spent many, many years practicing these techniques in order to acquire the total dynamic results. Of course, you are getting a scaled-down version. It is nonetheless beneficial, as well as powerful. These invaluable breath techniques are now yours if you choose to practice them. These techniques take practice to feel the full benefit and power behind them. However, once you get them, you will find many empowering ways to use them.

In the first stage in mastery of the breath, we learn how to maintain a holding pattern at the upward stroke of the indwelling breath.

By holding the breath at the completion of the indwelling breath, we create *Pranayama*. Pranayama is a term that describes the opening of all psychic faculties—"all channels open". Historically, this creates the ability to see through the "Third Eye; ("Brow Chakra" or "Ajna Center"). Holding the breath at this point is imperative because it allows for both the transference of oxygen and also, the formation of the intent of that which you desire (to do, be, or experience.) Holding your breath allows your intended thought to permeate your body. This completes this part of the process.

This "holding" of the breath is not "holding" as we are accustomed. It is not forced. Instead, it is a slight breathing in the uppermost part of your lungs. You will be breathing so slightly, that if someone were observing you, they would think that you were not breathing at all. This ever so slight of breath is called the *Micro-Breath*. This is the type of breathing that the yogis do in a deep state of meditation where they appear not to be breathing and are beyond the feeling of physical pain, beyond the physical realm with all of their psychic channels open. It is not something that takes a lifetime to learn. However, it does take some time and a good deal of practice. The first thing to learn is that doing the *Micro-Breath* will not cause a deficiency of oxygen. There is no need to fear not having enough oxygen. This is usually the first fear you will need to overcome. Again, it is not a matter of holding your breath; you will be able to maintain all body functions just fine. This technique of breathing will probably feel strange at first; it is just a matter of getting used to this ever-so-slight breathing.

Micro-Breath Technique

- Sit or stand comfortably with your spine perfectly straight.

- Fill your lungs completely as you inhale.

- As you breathe in, imagine that you are drawing in with your breath large amounts of prana and circulating this energy through all parts of your body.

- Now, let out 1/4 of your breath and begin to maintain the *Micro-Breath.*

- Your lungs will remain expanded, back straight, diaphragm in the upward position in your rib cage, rib cage completely expanded, and an ever -so-slight breath slowly going in and out of your lungs.

- You are not conscious of exhaling, for the air just seems to go out without noticeable movement of the lungs; it is a sensation of drawing air in on almost a continuous basis.

- You can also sense a slight movement at the top of the lungs and sometimes a coolness in the bronchial part of the lungs. Do the *Micro-Breath* again, and this time maintain it for as long as possible and feel this ever so slight in and out movement, even though your rib cage is fully expanded.

Many people can maintain the *Micro Breath* for as long as 20 minutes or more, and the yogis can do this type of *Micro-Breath* for hours. Now, try the *Micro-Breath* again, and this time focus your perception with your eyes closed from your Third Eye (or Ajna center), which is at the center of your forehead. What did you see? Light? When you maintain the *Micro-Breath* your alpha brainwave greatly increases, and all of your psychic capacities or channels begin to open. Again, this state is called *Pranayama,* whereby one is able to see through their Third Eye, Brow Chakra, or Ajna Center. Often, this sight is manifested in the ability to see light through the Third Eye. If this state is maintained long enough, you can have an out-of-body experience. The Kundalini also opens with the *Micro-Breath*. Kundalini is the life force or prana which comes in from the top of the head, and with the breath, goes down to the bottom of the spine in a rhythmic pumping action. When this energy hits the bottom of the spine (or base chakra), it spins up in waves in a form of a double helix (☿ Caduceus). This double helix of energy spins up and down generating an electrical field. When you draw in your breath you start the generation of this energy, and the *Micro-Breath*, (with the type of holding on the indwelling breath,) builds the magnitude of the electrical energy. This is what is called the raising of the Kundalini.

The Pulsed Breath

The *Pulsed Breath* is used for instant balancing and centering. It will bring equilibrium to our mental, emotional, and physical bodies for neutralizing negative or critical thoughts or patterns. Too much positive energy, such as anticipation or over-excitement, can also create imbalance which can be balanced with the pulsed breath. The *Pulsed Breath* is the carrier wave for our thoughts and the intentions behind them. To do the pulsed breath:

- Draw in the breath with the intention of that which you want to do.

- Hold the breath on the upward stroke of the indwelling breath, doing the *Micro Breath*. Focus your energy from the Heart and Brow energy centers of your body, onto your intention (that which you wish to do, be, or perform.)

- When you have reached the level of the greatest magnitude (or critical mass) the breath is quickly released through the nostrils with a strong expulsion and a snapping-type contraction of the abdominal muscles and diaphragm. The *Pulsed Breath* will sound and appear like a strong snorting out through the nose.

- With practice, the power of the *Pulsed Breath* increases.

Clearing and Balancing Technique

Using both the *Micro* and *Pulsed Breath*.

This is a technique to clear negativity, bring yourself to balance, and clear the ego-intellect out of the way for intuitive questioning.

This technique is so especially useful if you have been exposed to intense negative energy or if you feel out of balance. To clear:

- Draw in the breath with your intention to clear. Visualize love and light surrounding you.

- Maintain the *Micro-Breath*, build the energy level to critical mass; see the light building around your intention. Build up the energy, and then…

- Do the *Pulsed-Breath*. Feel your energy field and state of mind instantly shift to a state of mental and emotional balance. Now, you are ready to go about your day, or to begin using your tools of the Intuition.

The Intuition works effortlessly. A relaxed state of mind is essential for effective pendulum work, and the breath is the key for a balanced state of your mind. You will become more conscious of your breath. With continued practice and deeper and more meaningful breathing, you will experience more energy, vitality, and greater clarity of mind. The breath awareness techniques in this chapter will assimilate into your daily experience, if you practice them regularly. By using these techniques to clear negative patterns, and by replacing those negative patterns through affirmations and concentrated intent, you will bring an obvious increase to your quality of living. The breath is the key which unlocks the door to self-mastery.

Chapter 6

Pendulum Applications

We can use our Intuition to assist us in unlimited ways for everyday decision-making and intuitive problem-solving. We can use our intuitive tools to:

Choose: the right foods, diet, vitamins, minerals, supplements, crystals, books, remedies, careers, teachers, doctors, employees, employers, locations to live, the right house or car...

Find: the right job, lost objects, missing people or pets, buried treasure, underground water, gas, or electrical lines, valuables, underground water, springs, minerals, ore mines, metals, oil, directions, schools of fish, ruins, graves, old monuments and buried artifacts, tunnels, geo-energy (magnetic) lines...

Discover: relationship compatibility of partners, friends, potential employees or employers, personal motivators, probability of success for any situation...

Make: accurate business decisions, best consumer choices, career moves...

Determine: sex of unborn children or animals, plant needs and soil deficiencies, diagnostics for repairs of your house or automobile, weather conditions, written or printed errors, positioning antennae, whether foods are sprayed with pesticides, or irradiated, whether foods are organic or inorganic...

Test: seeds, soil, and potential planting areas, purity and quality of food and liquids, authenticity of signatures, nutritional value...

Analyze: physical, mental, emotional, or etheric body of either yourself or others— including all needs, deficiencies, imbalances, patterns, fears, or blockages.

We really *do* have access to what we call the Infinite Intelligence or the collective consciousness! This allows us access to the knowledge of our own innate wisdom, and also the collective consciousness of all humankind. We have the ability to access information and to find answers to any questions that we can imagine.

Again, the pendulum acts as the means to make what we already know in our Intuition available to our conscious mind. The pendulum has been used by people throughout history to determine an infinite amount of information and knowledge that could be considered "the unknown."

Everyone is a "natural" dowser, but some people have to suspend their innate disbelief before they can claim and develop this inherent intuitive ability. Continue your practice. If you get a response that is not precise, be patient; remember that you are still a student. Mistakes can go a long way to assist you in refining your dowsing skills. Remind yourself of the nature and specifics of the pendulum language and make the effort to understand why you may have received an inaccurate response. Look to the effect, then trace it backwards to the cause. The problem you are facing is not about your ability to dowse; it is about the refinement and clarity of the questions you formulate and the image you hold as you go through the dowsing process.

Now that you have developed good pendulum form, let us continue on with the journey in applying your intuitive skill to the unknowns.

Direction Finding Method

Follow the subsequent instructions to find: The direction of land, water, lost object or person, buried treasure, ruins...

- Stand erect with your pendulum suspended in the neutral, or search, position, out about 12 inches, directly in front of your chest with your elbow tucked against your side, Watch the pendulum out of your peripheral vision for any changes in the direction of the swing.

- Be prepared to turn your body clockwise in a complete 360 degree circle.

- Begin the swing of your pendulum in a sideways motion

- Fix the object or substance of your search firmly in your mind. Repeat it over and over, if necessary, to keep your focus.

- With your left arm stretched out, point with your left hand or finger.

- Slowly begin pivoting your body clockwise (right), repeating in your mind "the object of this search can be found directly in line with my pointing left finger?"

- Slowly continue your turn, repeat your question often, until you notice a deviation in the swing. This is a change from a sideways oscillation, to a maybe, greater-or-lesser position, and then into the forward and backward (or *yes*) swing.

- When you are positive of the change, inscribe a line in the dirt with your foot or place a marker, (a piece of wood, stone etc.) in alignment with the determined direction.

- Repeat this process as you slowly continue turning your body clockwise until you return to the beginning point. Throughout the search you may find one or more changes in the swing of your pendulum. It is possible to have two or more *find points*. This may be somewhat confusing. This may be due to a distraction, mind wandering, or in reality, it could be two find points. There are other possibilities that could be connected to your search in some way that could give you an indication of a *find point*. A piece of information or a fragment of the lost, buried, hidden object, though of secondary importance, can cause your pendulum to respond.

- To double-check your findings, return to your starting point. Now slowly pivoting your body counter-clockwise, repeat all the steps. Observe the pendulum's swing. Check to see if the pendulum's motion shift corresponds to your earlier clockwise pivot. If still you get confused answers and feel uncertain about the direction of the *find points*, begin again with the first step and repeat the procedure. Keep your attention focused on the objective of your questioning. Resist being distracted by physical movements and surrounding visual stimuli.

- To double-check your answer(s), align yourself with the directional marker(s), swing your pendulum in a search mode for a *'Yes'* or *'No'* response, with the question, "the object or substance of this search can be found in this direction?" If you receive a *Yes* answer, you may proceed with confidence in the direction of the object, or substance, of your search.

With practice, the coordination of your pendulum and body movements will become better, and finally, automatic. You can, with practice, master the technique of finding direction, resulting in an increase in your dowsing sensitivity, accuracy, and development.

At this point we have learned to determine the specific direction(s) that will lead us to the *find point* of the object or substance of the search. Since we don't have a map where "X" marks the spot, the next step in our development is to be able to locate the *find point*.

Locating Find Points

You have determined the direction(s) of the substance or object of your search. To locate your *find point*:

- Walk forward slowly with your pendulum swinging side-to-side. Repeat the question in your mind..."The *object* of my search can be found here?" Your pendulum language in this procedure calls for the instrument to change from the side-to-side (or *No*) swing to a *Maybe* swing position as you come into the vicinity of the *find point,* and then finally into a forwards-and backwards (or *Yes*) swing at the immediate edge or point of contact with the object or substance of the search. The change in swing will occur regardless of whether the object is overhead or underground, moving or stationary, disguised, mixed, fragmented, or scattered.

- The instant your pendulum indicates a change in swing, mark the spot. You are most likely at the *find point.* Some caution needs to be exerted. If you are walking too fast, you may not have given your pendulum enough time to alter its movement. If this is true, you may have walked several steps beyond the starting parameter(s) of the object, or substance, of your inquiry.

- You have now located the *find point.* If the parameters of the object of your search are vague, (for example, an underground stream or mineral vein), then you have found the boundary of one side. If this is the case, mark your determined edge of the object and walk several yards across it, turn around and approach the object from the opposite direction using the same method previously described.

- Repeat these steps from all directions, or from enough points, until you have a general outline of the *find point* area.

- To mark the edge for determining the full parameters of your underground stream, pipeline or cache; walk slowly, inscribing a line in the earth with a stick trailing beside you. Watch for any deviation of your pendulums' forward-and-backward swing. If the line is curving in its form, make notes of the arc or change in course as you follow it back to the starting point. If you start getting weak or erratic pendulum movement, it may be that you are getting off-track. Lay the stick down in line with the direction last noted; then, walk several steps away from, or perpendicular to, your marker. Now approach the stick (object's parameters) and watch for any change in swing of your pendulum. At the point of a *Yes* swing, you are once again on the true edge of the objects' parameters where you may resume your edge marking to completion.

Defining Depth

Now that you have learned to find direction, determine the *find point* or parameters of a location, defining depth of the object or substance is the next variable to establish. If your particular application requires measurement distance down or up to your *find point*, use your counting language, swinging your pendulum in the side-to-side (or *No*) motion. Ask for the depth to be indicated in inches, feet, yards, miles to the "top" of the object of the search. You can learn about the object's height by repeating your depth measurement, but this time, ask to gauge the bottom depth of the object or substance. When you are searching for underground water, mineral veins, etc., there are often additional streams, pools, or veins located below or next to the original *find point*. It is to your advantage to assess the area thoroughly to be certain that you are at the original *find point* before moving on.

Pendulum Applications

WATER

Finding pure water is an important issue, and getting more important daily.

The world community suffers from a lack of sufficient, unpolluted water. The water supply in the U.S. is polluted by toxic chemicals from agricultural run-off, manufacturing by-products, and deliberate chemical pollution. In rural America, some 15 million people suffer from a daily dose of unclean drinking water. Over 30 million Americans lack decent sanitation. In the developing world, 16,000 children under the age of five years of age die EACH DAY from water-related diseases. On a global basis, 30,000 deaths are reported daily as a result of water-related diseases. Dowsing can assist in the search for pure water, and can determine the degree of purity in any potable water. If you have developed your water dowsing ability and would like to assist in this critical world problem, contact the Global Water for Life organization in Washington, D.C..

Many beginning dowsing books encourage people to become instant dowsing experts. We do not want to give anyone the impression that dowsing for water is easy, and that anyone can do it. Dowsing for water is an art that each person possesses in varying degrees. It takes *practice* to reach your potential; there are some who seem to be born with this talent. We highly suggest that you spend some time in training with a competent water dowser before going out and predicting for others where a water well should be driven.

Finding pure water can seem complicated, and often there is a lot more to it than meets the eye. Like any other application, it is a matter of knowing how to ask the right questions. When a dowser requests potable water, he is requesting, by definition, water fit to drink. However, water fit to drink is not always beneficial. For instance, iron-rich water can be of value to those people who suffer from anemia, but it promotes cancer cell growth in cancer victims. Municipal water supplies are fit to drink, but we know that chlorine converts to the carcinogenic chloroform.

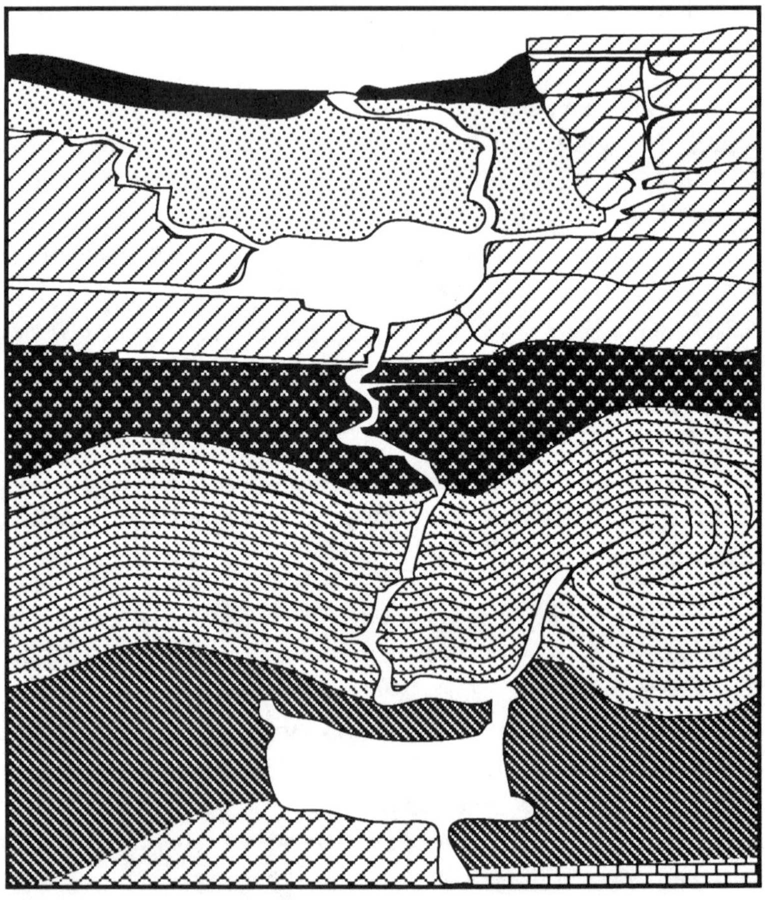

Water emerging from deep within the earth forming domes, or pockets of water that spread out into streams, or veins. The veins radiating outward from the water dome are the objective *find point* for a potential well-site. Sometimes these veins may reach the surface as pure spring water.

As we dowse to determine well sites, should we also be asking for long-range health benefits from the potential water wells? Will the water well be pure two hours from now, two days from now, two weeks from now?

Testing Water: To test water for purity, form the question in your mind of the object of this test:"Infinite, indicate if this water is for my highest and greatest benefit?" Hold the pendulum over a cup of water, if it gyrates positively (clockwise) the water is fit to drink. Counter-clockwise, indicates that the water is in some way polluted and not fit to drink.

The second part of the test is this: form the question, "Infinite, indicate to what degree is this water pure?" Start out with a side-to-side or *Maybe* swing with your pendulum and begin the counting language while thinking in terms of percentage. For example; 10, 20, 30, 40, 50, 60, 70, 80, 90 and the pendulum swing goes to the *Yes* position. Keep going to 100 and it begins to deviate back to *Maybe*. Go back to 90, and continue counting 91, 92, 93, 94 and again the swing starts to move toward the *Maybe* position, return to 93. In this case 93% is the purity level of the water. Try this now on your own tap or well water, then compare it to a variety of brand bottled water. Currently there are very few regulations on bottled water, and just because it's bottled water does not mean it's pure. We suggest that you even check out the differences between the bottled waters as to the percentage of purity.

Practice Test: Test a row of a dozen cups of water. Mix 1 teaspoon of salt into one cup. Now have someone mix the cups up so that the one containing the salt is even unknown to the tester. Check each cup for percentage of purity. This is a good test for development of accuracy. If you have trouble choosing the correct one it does not mean that you can't do it, it just means that you need more practice.

Finding Water Practice:

- First, swing your pendulum in the side-to-side motion and slowly walk toward the edge of a table or the back of a chair holding the *image* of finding water firmly in mind.

- As you get closer to the object representing the stream of water in your mind's eye, the pendulum will begin to swing to the *Maybe* (or more-or-less) position, and this may turn into a clockwise oval gyration. When you are directly over the search object, the pendulum movement will change to a full clockwise circle.

If you continue beyond the stream *find point*, the pendulum will deviate back to an oval and eventually return to a side-to-side or *No* position. Since you are making up the rules of communication here, you determine at what distance you want to be from the *find point* when the oval motion begins or ends, and then integrate that response as part of your pendulum language. Whatever response you develop, *you are the one in control,* so that any other response that is not part of your pendulum language may indicate either a random error, a distraction, or a clue that there is more involved than what you are seeing. If you experience conflicting results, repeat the test or try again at another time.

If you follow this procedure while maintaining a strong image of the object of the search and remind yourself of the response indicators, then eventually your responses will become automatic.

Food

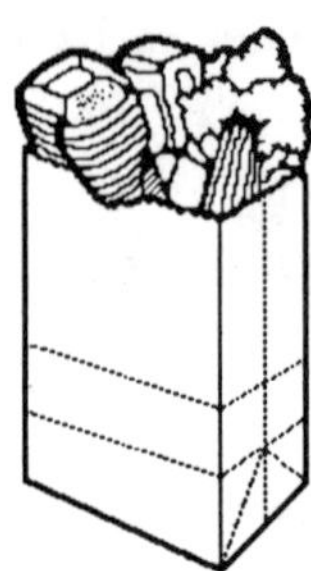

Food Testing: It is good to test your food before buying it in the store. You can determine the quality, whether the food has been sprayed with chemicals, or exposed to radiation, whether the food is organic or inorganic, or if you have an allergic reaction to that particular type of food. It is all dependent on your specific line of questioning. There are some factors in a store setting that could affect your results: fluorescent lighting, people, etc.. It may be best to do some of your more in-depth testing at home. However, it is easy to check for harmony or compatibility of your foods, vitamins, supplements, or anything else that you consume, with the following technique:

- Form your question in your mind and hold it there firmly: for example, "Infinite, indicate if this food is for my highest and greatest benefit?" If necessary, continuously repeat the question.

- Adjust your pendulum over the food and allow it to gyrate clockwise or counter-clockwise.

- Hold your left hand, palm down, between the pendulum and the food. For most people, the normal response of the pendulum to the back of your hand is usually a clockwise or positive gyration. If the pendulum continues to gyrate in a clockwise direction after you put your left hand over the food in question, it indicates that there is harmony between you and the food. Should the gyrations change direction from positive to negative, or clockwise to counter-clockwise, then that food is definitely *not* good for you, (or not good for you at that time). If the pendulum changes to a side-to-side oscillation, this also indicates that the food is not necessarily good or compatible for your needs.

We have witnessed this case in personal practice and have found such to be quite common: a woman with unexplainable headaches went to a doctor who could find no reason for them. The woman tried a pendulum over her food and found clockwise (good) for brown bread, boiled eggs, etc, and counter-clockwise for white bread, fried eggs, sugar, etc. She promptly changed her diet and to this day no longer has headaches.

It is very important to check your household products, your cosmetics and soaps, your cleansers, and anything with deodorants in it. Many of the products we use today contain harmful chemicals that can be the cause of allergies, rashes, headaches, or even low energy.

To test products such as foods, vitamins, herbs, and remedies in relationship to yourself, hold the object or substance in your left hand and suspend the pendulum over it. Again, the motions of the pendulum will be the same: clockwise is positive, compatible, harmony, or good for you; counter-clockwise is negative and not good, incompatible, or inharmonious for you.

When you compare two things that are closely related, you can gauge the difference by the degree or number of the gyrations. For example, you can compare two things, each positive. The greater the degree of movement, or the greater the number of gyrations, indicates a more positive answer or state.

Another way to determine differences between many potential items of choice is to ask the Intuition to assign a percentage of goodness that a particular product holds for you. This will allow you to test the differences between as many items as you would like, and to choose between items of close likeness. For example, when testing a shelf full of vitamins and minerals, you may come up with such levels of goodness ratings as, 55%, 65%, 80%, 85%, 90%. You may have one at 98%. This simplifies your comparison and minimizes guesswork. Use of the *Percentage Pendulum Chart* helps to make this process very quick and effective. When you are testing, keep in mind that your *answer is for the immediate time only.* Your results may be different at another time with regard to a particular food, so check often until you have established a consistent record. This is especially important when you test for food allergies.

Quantity & Time

Figuring Quantity & Time Factor: While you are holding the substance in your left hand, ask: "It is for my highest and greatest benefit to take _____, (for example, Vitamin C, 100 mg., 500 mg., 1000 mg., etc.) Now you will want to figure in the **Time Factor:** "It is best for me to take this once, twice, three, four, times a day, for one, two, three, four days, one week, two, three, one month, two, three," and so forth. It is not necessarily good to take vitamins too often; you could be telling your body to shut down its natural production of a particular vitamin. Sometimes, taking too much of a particular vitamin can be worse than not taking any at all. It is best to find out how much, how often and this method will help you to establish this information; "It is to my highest and greatest benefit to take this for, one, two, three weeks ON, and one, two weeks OFF, etc...

To check unknown quantity and time variables with a pendulum: hold the substance while counting the positive (or clockwise) gyrations. You can also use the backward-and-forward (or *Yes*) answer with the pendulum when counting: 1, 2, 3, 4. Perhaps when you get to 5, the swing will start going toward the *Maybe* direction. Then you back off to 4, and you see the swing return to *Yes*. It is good to recheck your answers often and to record the results.

It is a valuable skill to check all your foods, vitamins and supplements, (or anything else that you take into your body), and to be able to take the guesswork out of how much, how often. It is always easier to do a preventative maintenance plan than a cure, so a little time spent on this practice may well be worthwhile. It will prove very beneficial to both your quality of health and savings of your money. For further work with vitamins, minerals, quantity, and time factors, see the *Pendulum Charts.*

Analysis from the Human Body

Direct Analysis from the Human Body: Negative conditions or dis-ease in the body can be detected using the pendulum. These negative conditions can be at the level of organ, body part, or

system. By inquiring about the functioning capabilities of the body part in question, by formulating and asking the appropriate questions, the causative factors of the disorder can be determined.

We have the ability to detect any energy distortions or blockages that indicate imbalances within the human body. These imbalances have the potential to become dis-orders within the body, or may have caused disorders that already exist. Hold the pendulum over the body and form relevant questions. Begin with a neutral or forward-and-backward swing. Slowly move to different parts of the body and check for the degree of the total health in each area. The pendulum will respond to energy distortions or imbalances by either going to a *No* swing or to a counter-clockwise gyration. Pain in a particular area of the body may create a negative energy distortion in that area, but the source may be somewhere else. For example, a headache may be caused by a digestion problem. It is important to define through your questioning what level you are working on: physical, mental, emotional, energetic, or etheric. Often, we see that stuffed, blocked, or denied feelings, pain, trauma, or imprints are the cause of the symptoms or physical dis-orders. See *Pendulum Charts* for extensive analysis of the human body.

Fertility Cycle: Some women use the pendulum to establish times of highest and lowest fertility for either conception or birth control. This is all right as long as you are not influencing the answers by an unconscious desire to have either a child or sex. If you are going to try this, we would suggest you take a calendar and record your cycle objectively in advance. On the days that you have marked down as the ovulation time, pay close attention to any twinges or other sensations you may feel during ovulation; just keep a record and see how accurate you are with your fertility cycle forecasting. This method as a birth control device has an accuracy of greater than 80% for those who have used it. This accuracy rate is comparable to some birth control products currently on the market, and may not be so hard on the body. We suggest that this method be used only if you have had several months of accurate records so that you are sure about your results. Also be aware that your results may be biased due to unconscious desires. This method may be best administered by someone other than the woman in question; this can increase the objectivity of the test and consequently increase the accuracy.

Sex Determination: The sex of an unborn child or animal can easily be determined by using the pendulum. The *Yes* swing (clockwise) will indicate a male, the *No* swing (counter-clockwise) will indicate a female. In the case of multiple births, appropriate questioning will allow you to determine how many males and how many females are in an animal's litter. First, establish how many total animals there are. ("There are 1,2,3"). Then determine the number of males and females. Even the time of birth can accurately be determined by using the techniques described thus far in this book.

This application is currently being used at an egg ranch in Southern California. The pendulum is used over the eggs to determine whether the egg is fertile or not, and whether it contains a male or female chick. The egg production at this ranch has increased immensely. They found that the pendulum user at this ranch can do this type of testing for six hours or more with total accuracy. Point the egg to the North, suspend the pendulum above the egg and keep your other hand on the table near the egg. A longitudinal swing indicates sterile, clockwise gyration indicates male, and counter-clockwise indicates a female.

Animals

Animal Analysis: You can do analysis on animals as in the aforementioned procedure for humans. This method can include testing for feeds, medications, supplements, and also for health problems analysis. Use the pendulum to determine the energy blockages and imbalances, or to answer specific questions about a particular symptom or dis-order. Animals are less complicated than humans and will have no ego involvement to influence the pendulum results. In other words, animals are less apt to produce psychosomatic or placebo responses than humans.

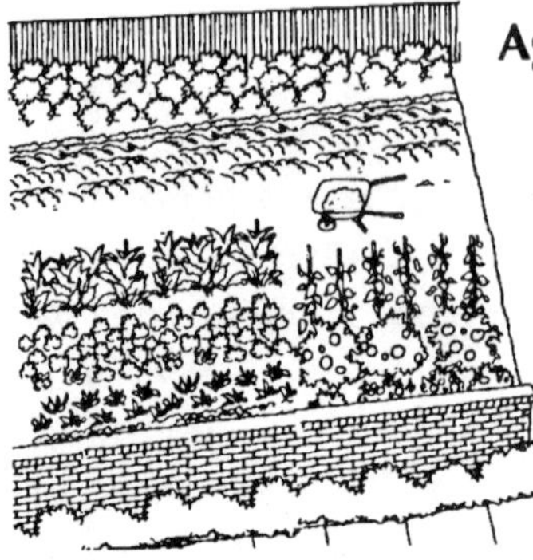

Agricultural

Agricultural: Use the pendulum and this technique to get a better return on your labor: plant the right plant in the appropriate place at the right time, and use the correct fertilizer or soil additives. The objective of this test is to determine whether harmony exists between the plant and the soil, and between the plant and the fertilizer. If a plant is not in harmony with the soil in which it has been planted, it will grow, but it will not thrive. Have you ever wondered why some of the trees or plants we plant do well, and others do not, while some just up and die? Many times we plant trees or crops where WE want them, either for looks or convenience. This may not be what is best for that tree or plant. Have you ever stopped and asked what the tree may want? We do well when we are in a place we want to be, even if the conditions are not perfect. Rarely are we well in a place we do NOT want to be, regardless of how perfect the conditions may be! So how do we know where the tree wants to be? ASK! That's the gift from the Intuition. We can ask about something we know nothing about and most often get the appropriate answer. For example; "Where, on this property, is the *best* location to plant this tree that will ensure its health, vigor, and productivity for the longest period of time." Use the *direction finding, find point method* and/or the *Direction Locator Pendulum Chart.* Once you find a suitable location, you can ask what diameter and depth the hole should be, and which direction the tree should be planted. As a tree grows naturally, it develops a "front door" through which it is believed that energy enters a tree. For the best results, the tree should be replanted with the front door facing in it's original direction. You can determine the front door of the tree by using the pendulum technique described in this section.

With the help of the Intuition, we can eliminate the guesswork and have a healthy, happy and abundant garden or crop.

Soil Test: On a piece of paper, place a small sample of the soil from the plot of ground that you are gardening (a large handful is sufficient). About 18" away from the pile of soil, place your plant or seedling. Hold the pendulum over the soil and watch for a strong gyration of the pendulum. Then suspend the pendulum over the plant and watch the gyrations closely. If the gyrations are clockwise and increase when the pendulum is placed over the plant, then the soil is in harmony with the plant, and is therefore suitable for growing. If the gyrations are clockwise and decrease, it means that the soil is not particularly good, although not necessarily bad for the purpose. Should the pendulum change to oscillations of forward-and-backward it signifies that the soil requires some form of fertilizer to make it suitable. If the pendulum gyrates the opposite direction when held over the plant, then the soil is unsuitable and no attempt should be made to use it. This test can be applied to anything that grows in the ground, but a living specimen must be used as a sample. If there is an indication that some sort of fertilizer is needed to compensate for some deficiencies, then try the following test.

Soil Nutrient Balance Test : Have samples of about two ounces of each of the fertilizers you may use. Place each substance several inches apart so that they are clear of each other's influence. Hold the pendulum over the plant. When it is gyrating clockwise, move it above each fertilizer and see which one gives the strongest reaction by the stronger gyrations or the greater number of gyrations. Now, determine how much is required. Slowly add fertilizer in small measured amounts to the heap of soil, and the gyrations will increase. When the maximum is reached the gyrations will decrease. Careful measurement will give exact proportions of fertilizer to soil; too much can also be detrimental. This method can be used for any nutrients that you may wish to use to supplement your plant kingdom.

Automobiles

Automobiles: This procedure has been very successful in diagnosing engine or car problems. Hold your pendulum over the engine, with the engine not running. While the pendulum is gyrating in a clockwise direction, begin to touch the various parts of the engine with your left hand: battery, valves, plugs, carburetor, etc. When a faulty part is touched, the pendulum will come to a stop, change to a *No* oscillation (side-to-side,) or to a counter-clockwise gyration. Most answers in this diagnostics will be in the more-or-less category; consequently, we suggest using the counting pendulum language with percentages to discern what degree a particular part functions.

There are some dowsers who tune cars with the pendulum as finely as electronic tune-up equipment. One such dowser, Marcel Trieau, has illustrated this talent many times at the American Society of Dowsers convention. He simply points his finger to various parts of the engine and asks his Intuition to indicate whether that part is in order or not. He can also tell if that particular part is in only "so-so" shape by the intensity of the rotations of the pendulum. Trieau is a skilled mechanic and is an example that a combination of a skilled intellect and the Intuition can solve complex calculations through past knowledge and experiences. In other words, he feeds in the proper information from the intellect in order to form the correct questions so that he is able to receive accurate results from his Intuition.

Generally, answers are greatly aided by your skill, experience, and the stored knowledge of your subconscious mind on the subject of the test. However, we have also seen how a person can set out without any prior knowledge regarding the subject and be successful in acquiring accurate results. Of course, the less knowledge you have on a subject may directly reflect the degree of flexibility or patience you will need while you stumble around finding the correct questions to ask.

Hazzard Detection

Noxious Radiations: underground water, electrical & magnetic appliances, electrical high power lines.

Just because we can not see some negative energy field does *not* mean it can not hurt us. This is a good example where our abilities to see the unknown can save us from undue pain and suffering.

It appears that underground water alone, or in combination with some known or unknown ray factor, can be harmful to some plants and animals. For example, certain trees will thrive in such areas. Oak, ash, alder, and willow do very well with underground streams. However, if you plant an apple, pear or other fruit tree in such a spot, it may wither and die. Some trees and plants seem to like the underground water and some don't. Animals are the same way. Cats have no objection to sleeping over a stream or so called irritation zone. Ants make their home over an intersection of two of these streams. Those beehives that house much stronger colonies than others are always over a stream of underground water (The American Dowser, Volume 28, No.2). Yet other animals stay clear of subterranean water. Dogs will sleep out in the open on a wet winter night rather than in a warm kennel that is over underground water. Pigs prefer to take a new born litter out into the cold rather than allow them to stay indoors over underground water. Cows seem very sensitive to these irritation zones and seem to produce less milk as a result.

In addition to these factors, there seems to be ample evidence that these elements may be affecting the health of human beings. Even though there is a lack of scientific agreement, the available evidence regarding this correlation between underground streams and adverse effects on the human body is compelling. Research in foreign countries has pointed to a strong correlation between various illness and the existence of certain earth energy forces from underground running water. Experiments by Dr. Joseph Kopp have suggested that above these subterranean streams and underground water currents several abnormal physical conditions will be found. These include magnetic anomalies, an increase in electrical conductivity of the soil and air, acoustic changes, an increase in the field strength of UHF waves, and an increase in the intensity of infrared radiation. Currently, there is no certainty or understanding of how these underground forces relate to the negative effects that may be produced. Still, enough research has been conducted to suggest that people should avoid places where a dowser gets a negative reaction either from underground water or other suspected underground factors.

Kathe Bachler, in her best seller book, <u>Discoveries of a Dowser</u>, tested 500 cases of people with either malignant or benign tumors and found, without exception, that these people slept over strong earth radiations. Chronic troubles such as arthritis, rheumatism, multiple sclerosis, and asthma are all strong indications of exposure to strong earth radiations. This does not necessarily mean that geopathogenic zones, or their magnetic effect, produce cancer, but it is possible that coupled with an additional antagonist, cancer may be more likely to form under such conditions. Noxious earth radiations can debilitate the human organism by reducing it's immunity level.

Pathological effects of sleeping over underground water may take years to become apparent in adults. But in the case of children, we can see almost immediate effects if they are susceptible to irritation zones. This was demonstrated by Dr. E. Hartman. He put a small baby in a cart which he rolled over an irritation zone. The baby started to cry, but stopped when past the irritation zone. This experiment was repeated several times with the same results.

The energy around the places where we spend a great deal of time can have a profound impact on our health. This is especially true of the sick or disabled who may spend a great deal of time in one place because of their health. (For example, in their beds or chair.) Since geopathogenic zones can have a deleterious effect, it is especially important that the ill person find a healthy place to live and sleep.

Electro-Magnetic Test: for harmful or noxious radiations from: televisions, computer monitors, microwave ovens...

You will find it interesting to test your television set or microwave oven to see how far noxious radiations from them reach into your environment. With your pendulum, start as far back as possible and slowly walk toward the object you are testing. Watch where the movement of the pendulum changes to a negative or counter-clockwise gyration. Then check from the side of the object. With the television on, I have found the radiations to be as far-reaching as six to ten feet, and with microwave ovens, three to four feet. It is advisable not to let children, animals, or yourself, sit within that negative field. Over time, these noxious radiations can have an adverse effect; they can weaken the immune system and be a factor in causing health disorders.

Experiments in recent years have brought to light many possible links between electromagnetic fields and increased childhood cancer and leukemia rates. Electromagnetic fields are intangible auras of energy which emanate from all electrical devices, from the biggest power line to the lowliest two-slice toaster. They are composed of two intimately related, but distinct types of energy fields: electric and magnetic. A power line radiates outward to the environment by means of these fields of influence it produces.

It now appears that magnetic, rather than electric fields, are most intimately associated with cancer. Normal cellular functioning and development may be disrupted by magnetic pulses.

Small lines carry more current and less voltage than the big lines, making their magnetic fields relatively stronger. While running at only

120 or 240 volts, these lines carry currents higher than the largest power lines. We live in close proximity to these small lines, but are usually quite a distance from high power electrical lines. This is not to say that the high power electrical lines are safe; being in relative proximity to them is very stressful on the human body. We have heard of adverse health problems directly related to living too close to high power electrical lines. High power electrical lines are believed to create direct interference with the central nervous system, and can cause neuromuscular confusion, disorientation, and abnormal distribution and absorption of minerals. In our testing with overhead electrical power lines, we have found that adverse effects on the human body can be felt as close as: thirty-five to forty feet for 2,000 volt lines, three-hundred forty to three hundred fifty feet for 230,000 volt lines, etc. You may want to check out the noxious radiation range of overhead power lines in proximity to a house before renting or buying.

Chemical Health Hazards: Using a dowsing analysis method can reveal chemical health hazards or problems which will not be found by conventional medical diagnosis. Micro pollution has been discovered by analytical techniques including gas chromatography, and mass spectrography techniques. Researchers can now analyze contamination in water and food down to parts per billion and parts per trillion. They are talking about pesticides, herbicides, toxic industrial chemicals, heavy metals, low level radiation, and also things like microwave radiation that can now be detected in our foods and water. All of these harmful substances seem to behave in the same manner on the biological organism. At very low levels they are toxic: they disturb the enzymatic functions, block liver and kidneys, and create serious illness. In general, they create a dis-order called immuno-suppression of the immune system. This disorder can render the immune system weak and ineffective, and it opens one up to a host of dis-eases. Examples of such immune system related problems include asthma, allergies, lupus, multiple sclerosis, and arthritis, among others. It is because of such problems with worldwide pollutants showing up in our foods and water that I feel so compelled to teach people how to determine what is and what is not for their highest and greatest benefit. As with most of the previous applications, the counting-percentage method for measuring the degree of goodness can be quite effective. See *Pendulum Charts.*

Business

Business Decisions: On the job and in business, using the Intuition can give you a great advantage over the normal trial-and-error approach to decision-making. If you are wavering  between two alternatives in a business decision, you can use your Intuition, pendulum, *Yes/No* language and/or *Percentage Charts* to find the best course of action. Using the Intuition is an excellent way to help give you a boost in making up your mind regarding appropriate choices. Additional applications would be to use the Intuition in locating a specific area or item such as a building site for a business, the perfect store front, the right tool or truck, etc.

A very effective use of the Intuition in business would be for determining the best employee for a particular job. Finding the best employee to hire can be very tedious, time consuming, and, the best employee may not always be apparent from the application. This task can be greatly facilitated by taking your stack of applications and resumes, reading the name of the applicant, and scanning the pages of the application. Put your left (receiving) hand over the signature or photo, use your pendulum and *Percentage Chart*, determine to what degree this applicant will be best suited in filling the open position, assign a percentage to this individual, and then move on to the next applicant. We have found this to be very effective and accurate, especially when there are many closely qualified candidates and the determining factor cannot be gauged from the application.

The Intuition can also be used in sales to locate prospects and prospective areas, or in land deals where it may be necessary to locate water or minerals. One of the most effective uses of the pendulum has been in telephone conversations where a snap decision needs to be made, or in understanding what the person on the other end of the line is trying to say, or in determining the validity or truth of what is being said.

Using your Intuition in weighing the pros and cons of any situation before making a business decision can save you a great deal of time, effort, and money. The more you use and develop your Intuition, the more you will find ways to use it in business or personal life applications. This can result in greater quality and success in your life.

Years ago, I had a construction company and needed to make contracting bids on a regular basis. For many in the business, this was a procedure that could take one day to one or more weeks of detailed planning, telephoning, and footwork to determine the total figure to submit to the customer. In most cases, if the contractor did not get the job, most, if not all, of the time required to make the bid was uncompensated. After having been in the trades for quite a few years, I did have a fairly good idea as to the cost of things; nonetheless, I didn't like all the time it took for figuring bids and estimates. I started to use my Intuition and pendulum to determine the right cost figure of the job for both myself and the customer. After checking out the job site and talking with the potential customer, I would go out to my van and come up with the appropriate estimate. At first, to determine the right figure, I would ask myself: "How badly do I need this job?", "How tough is the competition?", "Can the customer afford what their ideas will cost?" However, I found that this type of questioning could be somewhat fear-laden, and that could have a negative effect on the results of the test.

A more inclusive type of questioning that I found to give the best results was; "It would be to my highest and greatest benefit to take on this job with _____________ (customer name)?" If *Yes*, then, "Infinite, indicate the perfect cost figure of this job for ___________ (customer) that will be for the highest and greatest benefit of ___________ (customer) and myself?" This procedure would take me a whole five minutes, and to my customer's surprise I would come back with the bid or estimate in hand. The interesting thing about this was, first of all, I had only really great customers. Secondly, I always got the job. The surprised looks on the faces of my customers were always fun, for typically my figures were directly aligned with theirs. Comments like; "Well, that's exactly what I had in mind, when can you start?", were quite normal.

Sometimes the jobs ended up taking more time or more money, but, usually the customers were great about it and balanced out the overrun cost. As the result of using my Intuition, I know without doubt that I made better decisions in a fraction of the time and saved both my customers and myself a great deal of time and money with a minimum amount of stress. The Intuition was also incredibly helpful in on-the-job trouble shooting, usually leaving the customers and the other tradespeople amazed as to how I was able to solve the problem so quickly, even with many unknown factors.

Treasure

Finding Treasure: Almost all treasure hunters seem to have one main problem in common. They can do the research—rummage through dusty old records and newspapers and stale smelling court records, find out who did what to, and with whom, and work out accurately the best location where the treasure is most likely to be found. However, when they finally arrive on the scene, most often, they still cannot dowse exactly the right spot; they cannot just march up to one spot and say, "Dig here!," and after digging, find the treasure! They can find the general location without difficulty, but NOT the exact location! We have seen this to be true on numerous occasions. There are many reasons for such apparent failure. There may be powerful *Thought Forms* surrounding the buried treasure. Influential energy of those who hid the treasure may, in a sense, still be guarding it. There seem to be many things and events that can confuse the area of the search, making it next to impossible to find the exact spot or *find point*. I have heard myself saying quite often, "Oh well, if it were easy, somebody would have already found it." Before you go out and attempt dowsing the location of some hidden treasure, you should first teach yourself to find hidden objects. Once this is accomplished, have someone 'dirty up' the location with artificial *Thought Forms*, and try again. Anyone who can learn to do this on a consistent basis will greatly enhance their chances for finding the real thing.

> *With practice and common sense, you will learn what the pendulum (Intuition) will and will not answer for you. Much will depend on where your limitations are in your thinking or belief systems at any particular time or place. Your only limitations are the ones you believe in.*

Chapter 7

Pendulum Charts

The *Pendulum Charts* are very effective in keeping the intellect or ego entranced long enough to allow the Intuition to be completely present with little interference from the rational mind. Additionally, the charts allow the Intuition to communicate about the subject matter through questions, in ways that you may have not thought about asking. The charts will give you numerous ideas about different things to look for, and different ways to ask pertinent questions. *Sometimes you can learn more from the questions than you will from the answers.* These pendulum charts are "user friendly"; the more you work with them, the more you will discover how to use them, and the more you will learn how to adapt them to many other types of applications. Use the *Knowing Your Intuitive Mind Pendulum Chart Applications*: as the software interface between the Intuition and the conscious mind.

When you work with these charts, remember to do the threefold permission before you start, especially if it involves others. It is imperative that you have permission and that you determine the appropriateness of your line of questioning.

You will discover that the applications are infinite for finding answers about yourself or others. The *witness* method described herein has proven to be a quick and accurate means to access your intuitive information and knowledge.

Each chart is described, and you will read about lines of questioning to consider as you attempt to uncover information. Remember, this is a dynamic and interactive process that will require your focused attention and creativity. As always, your answers will be only as good as your questions.

Master Chart

Master Chart: This is the menu for starting. This chart will assist you in beginning your line of questioning with a particular situation, dilemma, or challenge. This is the lead-in to all other charts when you have no clues about a starting point. When you get toward the end of a line of questioning, or come to a blank state of mind, it is good to refer back to this *Master Chart* and ask if there are any other charts that would be helpful in a wholistic approach to the particular situation in question. Remember to keep in mind all variables, including your state of mind, energy level, and the proper time and place.

Pendulum Language Chart

Pendulum Language: If you haven't already established your communication line with your Intuition, then this is the beginning point for you. Read the section on establishing the pendulum language (Chapter 2), and the basic instruction enclosed with the pendulum charts. Review *Yes/No.*, forward-and-backward, side-to-side, clockwise and counter-clockwise language. *Maybe:* is more-or-less degrees of swing between the *Yes* and *No.* position. If you get no motion, or erratic motion, try again at another time or rephrase the question.

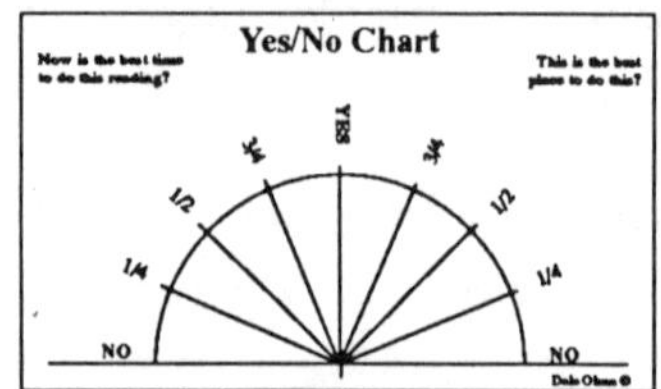

Yes/No Chart

Yes/No Chart: This is your most commonly used chart in combination with other charts. Before starting your line of questioning it is best to get in the habit of asking: "Now is the best time to do this reading or questioning?", and "This is the best place to do this questioning?", and the threefold permission. This is a great chart to determine answers that are not just *Yes* or *No*. The answers to most questions will fall into the category of *Maybe* or more-or-less. Most often you will be using this chart in conjunction with other charts.

Percentage-Probability Chart

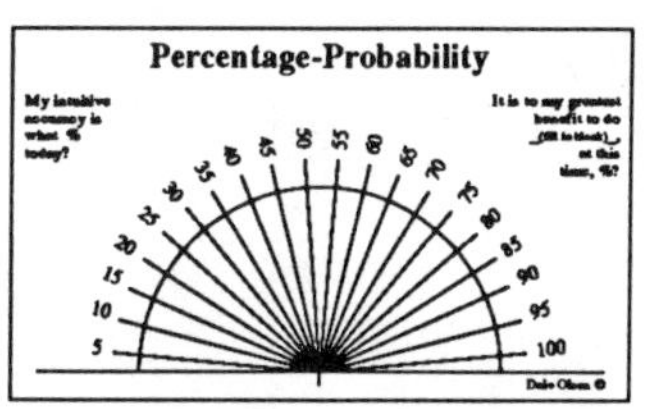

Percentage-Probability: Along with your **Yes/No Chart,** this will be the most widely used chart for comparing two or more like products, variables, or choices. You may find that ascribing percentages to various factors will be most helpful in determining differences in choices that may seem otherwise ambiguous. In other words, you can compare multiple variables and assign percentages to each, thereby putting your choices on a scale of most beneficial to least beneficial. In this way, problems can be weighed and choices made much easier, based on Intuition rather than trial-and-error. The main factor to keep in mind is how you word your questions to the Intuition. Be clear and precise, measure one variable at a time, and remember the time factor; what was true yesterday may not be so today. After using the **Yes/No Chart** to determine if time and place is appropriate for the test, it is then beneficial to use the **Percentage-Probability Chart** to ask, "My intuitive accuracy is what ____% today?" Then you may proceed with your questioning.   "It is to my highest and greatest benefit to ____________ at this time ___%?" This chart can be used in conjunction with all other charts in measuring or weighing any variable or combination of factors or choices. This chart is also very useful in determining the probability of success for a particular situation, event, or experience. You will find hundreds of ways to utilize this chart for just about anything you wish to measure.

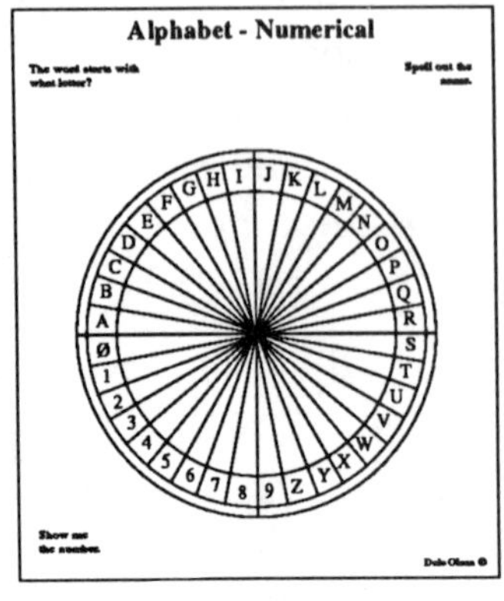

Alphabet-Numerical Chart

Alphabet-Numerical: This chart is to assist in spelling out a name, word, or number unknown to the conscious mind. This can be helpful in spelling out the names of your guides, teachers, or the names of places and things within the subconscious or collective consciousness. This chart can also be used in determining a sequence of numbers.

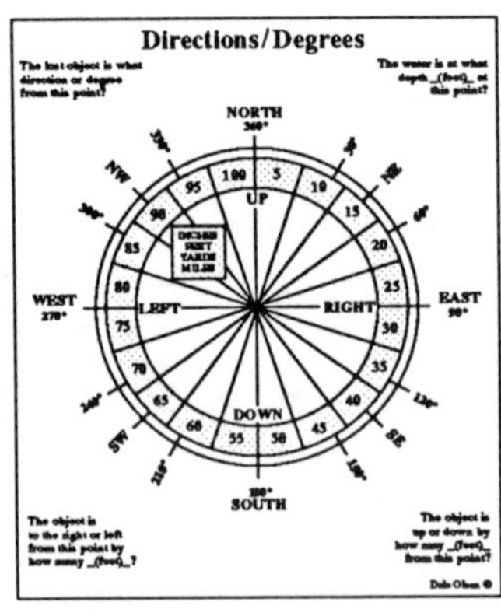

Directions-Degrees-Depth Chart

Directions-Degrees-Depth: This a three-dimensional chart. You can use this chart for finding a lost person, pet, object, buried treasure, archaeological site, water, oil, or minerals (such as crystals, gold, silver, etc.). What you are looking for can be miles away or hundreds of feet beneath the surface of the ground. It is still within our intuitive abilities to find them. This may be difficult to comprehend, and it may not be acceptable to the logical part of the brain. Nonetheless, it has been clearly evident throughout history that people have found water, gold, lost people, and many other things by using the Intuition.

A substantiated example of this ability is with the oil companies. Most of the leading oil companies in the world employ people to do what is referred to as "map dowsing" with a pendulum. Some of the major oil finds in the world today have been discovered through this method. (For further details on map dowsing, see chapter eight on "distant dowsing".)

All people, all substances on this planet, radiate at a certain energetic frequency or set of frequencies. Dowsing is a means of tuning into that frequency within a certain area. By using your intention and focusing your mind, you tune into the parameters of the frequency of that particular person, lost object, or buried treasure.

Use this chart in conjunction with a map. Orientate your map and this chart to magnetic North (use a compass). Begin by placing a dot on the map at a location within the *search* area. Place the dot on the map at a familiar landmark (post, surveyor marker, river, rock formation, etc.). Now you can use your chart: "Infinite, the lost object is what direction or what degree from this point?", or, "The object is to the right or left from this point by how many _____inches, feet, yards, miles ?", or "The water is at what depth ______feet at this point ?", or, "The object is up or down by how many ______feet from this point?", and so forth.

There is one note of caution on searching for lost people or animals. The person or animal you are looking for may not consider themselves lost, and, they may not want to be found. This could have a major effect on the accuracy of your results.

It will take some practice to develop the ability to locate lost or buried objects or substances. We suggest that you start with things that you can find the answer to (such as known buried water or electric lines,) or practice by having someone bury an object (like a ring or something to which you are attached.) Then go and locate it. Start simply and build slowly to develop this skill; starting too big can be very discouraging. Remember that this is a learning process, so build on your skills and practice.

Personal Motivators Chart

Personal Motivators: This chart helps you to run checks and balances on your emotions. This can determine both positive and negative aspects and can pinpoint emotional imbalance. By using the percentage chart in conjunction with this chart, you can determine to what degree a particular personal motivator prevails in your life at the present. If there is some question as to why you want to do a particular thing or what is behind making a particular choice: "The present issue evolves around which personal motivator ___________, to what degree ___(%)?". This will indicate very quickly whether your decision is based on a true positive emotion or out of fear or ego.

Relationship Compatibility Chart

Relationship Compatibility: This chart, in conjunction with the percentage chart, can be used to determine what level of compatibility exists with the people in your life. You can use this chart in a myriad of ways to measure or compare the aspects and attributes of a relationship between you and another person (your mate, children, family, friend, partner, employer, employee, teacher, physician, therapist, and so forth).

With the *Relationship Compatibility Chart*, begin your line of questioning with, "This person is compatible with me in what way ___________ ?", and visa versa, "Indicate what other variables are involved in this relationship with ____________________ ?," and, "To what degree is this variable part of the relationship ____%".

Sometimes it can be tricky to keep your objectivity when checking out others in relationship to yourself. Doing this type of inquiring can assist you in being clear about your own desires or ulterior motives toward the person in question. The results can be very interesting, for the answers can indicate something quite different than what you may have thought was true.

This chart has many features. It can be used for determining the possible longevity of relationships. Remember, any time you start questioning about the future, you stand the chance of being incorrect. Time factors are the hardest to predict because you or the person involved can change your mind(s) at any moment, thereby changing the end results. Another drawback to figuring longevity of relationships is that the future plans may not yet be established in the other person's mind. Whatever results you get, it is best to keep it in the framework of, "Well, this is very interesting however, we'll see."

In other words, keep this questioning light and easy, and do not take it too seriously, especially at the outset. If you are getting some negative results about the individual in question day after day, and you really feel that the results are from the truth of your inner knowingness, then respond in a manner that is true to your self and to your highest benefit. It is also very beneficial to complete your line of questioning by asking; "I am in this relationship because there are some very valuable lessons for me to learn about my _______________(self, life, strength, personal power, character, others, etc)?"

We interact with other people on four levels; physical, mental, emotional, and spiritual. Consequently, compatibility is more complex than it may seem because we must compare compatibility on each of these four levels. Questioning involving this area has depth beyond our normal linear perspective. This is important because you cannot look at the whole issue of compatibility without considering each of these four levels. Your questions must be specific in regard to these distinctly different, yet sometimes overlapping levels of existence.

Within your line of questioning, it is best to consider your compatibility to the other person, and also the other person's compatibility to you on each of these levels. This is where you can work with the *Percentage Chart* to weigh the various aspects and attributes within the relationship. For example, the line of questioning would be; "I am compatible with this person in what way ___________ (aspect or attribute) on what level ___________(physical, mental, emotional, spiritual) to what degree _____ %" (use *Percentage Chart*)", and, _______________(name) is compatible with me in what way _________________ (attribute/aspect) on what level _________________ (physical, mental, emotional, spiritual) to what degree _____ %." By following this line of questioning, you may become aware of what aspects and attributes have contributed to your attraction to the other person. You may also understand the differences in compatibility on each of the four levels, and how you both are focused toward each other.

Time Factor Chart

Time Factor: The Time Chart needs to be used in conjunction with other charts to determine the appropriate time factor involved with the use of vitamins, minerals, remedies, therapy treatments, exercise program, food supplements, need for particular foods, herbs, fertilizer, etc. For example, "This vitamin is needed how often ?", and then determine whether it is to be taken on a continuous basis or how many weeks On and how many weeks Off. Sometimes a program with a certain number of weeks On and a number of weeks Off may prove to be the most beneficial. Too much of a good thing can sometimes be worse than nothing at all. Be sure you determine the correct time factor in what you are measuring to be for your highest and greatest benefit.

Quantity Factor

Quantity Factor: This is similar to the time factor chart in that it can be used in conjunction with anything that has volume or mass to it, anything that requires an unknown measurement that is most appropriate for the individual or application. This chart can be useful for determining the quantity of vitamins, minerals, supplements, herbs, remedies, fertilizer, etc. The line of questioning would be, "The most appropriate need for this _________ is in what quantity __________, and how often __________?"

Plant-Soil Chart

Plant-Soil: As mentioned earlier, you can make a plant grow under some pretty adverse conditions, but you cannot make it thrive. Most people expect that all a plant needs is to be planted and it is going to do great. When the plant hardly comes to blossom, or produces very little fruit, they are greatly disappointed. This is the unfortunate result of the trial-and-error approach. By tuning into the plant, and by your intention alone directing your communications to the life force within the plant, you will have great success in determining exactly what the plant needs for optimum production. The *Plant-Soil Chart* can show you exactly what is needed in soil conditions (pH scale), fertilizer, water, light, etc.. Try the pH scale to determine whether the soil is acidic or alkaline by holding your left hand over a sample of soil while using your pendulum with the pH scale. If you have a chemical pH tester, run another physical pH test and compare the two. You may find that the two pH tests will be the same or very close. You can also at any time determine whether the plant or soil is deficient, balanced, or in excess of anything.

Wholistic Health Charts

Wholistic Health Charts: This is another menu chart similar to the Master Chart. It is directed toward a wholistic analysis of most aspects of the human mental, emotional, energetic, and physical body. *When in doubt check it out*, and you can start your line of questioning by using this chart: "The best chart to start with is _______________?", "The best chart to use next is ______?", and so forth. The charts in this section span many different belief systems and schools of thought. You will find that the more you work with these charts, the more you will find that the flow of your questioning will evolve into your own system.

Condition Chart

Condition: This chart can be used at the start of a test to determine the degree to which a particular problem exists, the condition of a particular organ or area of the body, or the overall state of health of the individual being tested. The chart is designed to handle many different physical, mental, or emotional conditions. For example, you can determine condition of blood pressure or temperature, degree of dysfunction, disease, or disorder, or the activity level of glands or organs. The line of questioning would be something like; "This

______________ (individual, dysfunction, disorder, disease, organ, gland, etc.) is exhibiting what condition ______________, to what degree _____%?"

Source of Condition Chart

Source of Condition: This chart covers the source or causation behind most of the existing symptoms that an individual can manifest. This is not to be interpreted as the final word about a particular dis-order or dis-ease; however, it does help to find the doorway through which one may approach the remedy that will be most effective. You may find there are more sources of conditions that could be added to the *Source of Condition Chart.* You may also have other sources of conditions based on different belief systems. These you will uncover yourself.

The line of questioning would be: "This present condition of____________ (type of condition) with ______________(name of individual) is from what source ________________, and to what degree of___(%)?"

Origin of the Dis-ease Chart

Origin of the Dis-ease: This chart indicates in what physical part of the body the present condition or symptom began. You can also use it for finding in what body part the present condition exists, or where there may be potential for problems in the future. In many cases, the results may point to where an individual has stuffed emotions and/or imprints from past traumatic experiences or from accidents that caused physical trauma. This type of imprinting from trauma may not always be seen on the physical level as problematic symptoms; however, we have seen hundreds of cases where such energetic imprints, in time, turned into dis-order or dis-ease. The line of questioning would be: "The (disease, disorder, symptom, origin, potential problem, etc...) is located where in the body _________, and it is (physical, mental, emotional, energetic) ________________ (Yes/No), and to what degree____(%)?"

Healing Remedies

Healing Remedies: The Healing Remedy Chart covers methods from conventional to alternative forms of healing. You may find that there are some forms of healing that are not listed and can be added. It is best to make this chart span the range of healing forms within your own belief system. Even if you may not be completely familiar with some of the listed modes of healing for yourself, they may be totally effective and useful for someone else. We have found that usually more than one type of healing modality is required. The line of questioning would start with: "For total health and well-being of ____________(individual's name), is in need of what remedy _______________, for this ___________(dysfunction, disorder, disease) to what degree ____(%)?", and, "This body is in need of what other remedy _________________, to what degree ____(%)?"

Weighing the choices by means of percentage gives you a means of comparing two or more types of healing modalities. This will enable you to make intuitive choices to assist you in deciding which is the most effective, safe, and least traumatic approach. It will also help to determine the appropriate order of the therapies. Adding the time factor to the acquired information will give an additional dimension: To find both the most effective means for the healing process, and also when it would be best to start or end one particular type of therapy. This "fine tuning", will determine the most effective form of therapy at precisely the perfect time. This will cause the least amount of trauma, have longer or permanent healing results in a shorter time period, and will result in less financial cost.

Systems of the Body Chart

Systems of the Body: This chart helps to give an overview to determine what systems are involved with the dysfunction or dis-ease.

The line of questioning is: "The system in need of attention is
________________, to what degree ____(%)?" Sometimes what may
appear to be a symptom originating from one part of the body may in
fact have its origin in another area of the body or system. This chart
can be most beneficial in determining potential dysfunction or dis-
orders. Being able to weigh each system with percentages can be a
great asset in a preventative maintenance program for yourself or
others.

Glandular (Endocrine) System Chart

Glandular System: This chart is based on the glands of the endocrine
system which control most body functions through chemicals called
hormones. The activity of these glands can have a drastic influence
on the health or balance of the body, and the mental or emotional
balance of the mind. The line of questioning would be: "Infinite,
indicate which of these glands are functioning improperly,
________________, and to what degree ____(%)?", or for an overall type
of testing, "To what degree is the gland ________________ (go through
all glands one at a time) functioning properly ____(%)?"

Chakra System Chart

Chakra System: The chakras are energy centers that are connected
within and throughout the energy system of the body. These energy
centers were identified in ancient times, and we now have the
technological equipment to measure these centers and verify their
existence. Some people are able to physically see these energetic
centers of the body. (This is called 'expanded sight').

With this chart you can determine how open or closed a particular
chakra is. This indicates how well the energy is flowing to a particular
area or how much blockage there is in that area. This will determine
the vitality or total health of the organs within that area. This chart will
assist you in measuring the chakras without having a developed
expanded sight. You can determine which chakras are not functioning

properly by having the Intuition indicate to what degree these centers are open, whether the energy is deficient, balanced, or in excess.

The line of questioning with this chart would be: "Infinite, indicate the Chakra in need of work __________?", or "Indicate how open this chakra is______?", "Indicate the degree of balance of this particular chakra _______?" We have found it to be very beneficial to measure someone's chakras before performing a form of therapy and then measuring them afterwards to determine the effectiveness of that particular therapy. The benefit or effectiveness of the therapy will show up immediately within the chakras. This is a great form of feedback that can be quite beneficial for the client. To visually see an actual improvement in measurement through the chakras will assist in the integration of that improvement into their belief system. *What you believe will be* is always the actual reality. There is a great deal more that can be learned about the chakra system. We have seen it to be an incredibly accurate and dynamic way of determining the total health of the physical, mental, emotional, and etheric states of being. In years to come, we feel that analysis of the chakra system will become the main form of diagnostics of the human body. For an extensive and accurate view on the chakra system, we would suggest reading "Wheels of Life" by Anodea Judith.

Color Radiation & Color Need Chart

Color Radiation & Color Need: This chart corresponds with the *Chakra System Chart*, and, is another way of indicating to what degree the body or chakra is deficient in a particular energy/color. In this case, light and color radiating at certain frequencies is considered to be a necessity and is utilized by the body to remain vital, balanced, and healthy. The origin of this perspective is from ancient times; nonetheless, metaphysics and today's science are now beginning to accept that color and light is a vital part in the health and wellness of the human being. The line of questioning for this chart would be; " Infinite, this body ______ is deficient in what color ________to what degree _______ (1-9 scale or %)?", and, "This body ________ is in excess of what color _________, to what degree ____(1-9 scale or %)?"

Nutrition & Food Allergies Chart

Nutrition & Food Allergies: This chart is used to determine where there may be a deficiency or excess in one or more of the major food groups. The most widespread use of this chart is to find what food allergies exist in what major food groups. Once that is determined, you can make your own list of all the food types within that major group consumed by the individual being tested. From there you can find the foods that are responsible for allergic reactions. This can be accomplished with either a Yes/No or a percentage answer to your list of potential allergenic foods. The line of questioning would be: "Infinite, the body of ______________(name) is in need of what food(s) _________(food group, specific types), to what degree ____(%)?", "This body _____________(name) is in excess of what food(s) ___________(food group, specific types), to what degree ____(%)?"or "This body _____________(name) has allergic reactions to what food(s) ___________(food group, specific types), to what degree _____(%)?"

Food Supplement Chart

Food Supplements: This chart can be helpful to those who may be in need of supplements to their ordinary diet. Sometimes these may be appropriate for chronic conditions or for those who find today's foods lacking in the necessary nutrition for total health and well being. There are more food supplements that could be added to this list and for some of you, there are some that could be eliminated. Again, we suggest that you change this chart to correspond to what works for you. The line of questioning for this chart would be: "Infinite, this body _____________ (name) is in need of what nutritional supplement ________, to what degree____(%)?"

Vitamin Chart

Vitamin Chart: Choosing the correct vitamins, for your own needs or for others, can be accomplished with this chart. Most people choose their vitamins by guessing or because someone told them it would be good for them, or because they read something about a particular vitamin and therefore feel that they need it. Sometimes we may even be correct when we think we might need a certain vitamin. Often the body will yell loudly, so to speak, to let us know we have a need. Sometimes we actually even hear it.

It is a well-established fact that the vitamin and mineral content of our foods is much less than it was many years ago. Consequently, we find ourselves needing to supplement our diets to fulfill the requirements of our demanding lifestyles. Using this chart to fine-tune your need for a particular vitamin can be of great value to you. Using the typical "shotgun method" of vitamin-taking can be detrimental; too much of a good thing can be just as harmful as not taking anything all. Selecting the correct vitamins at the right time, and taking them for the appropriate duration can make a world of difference in the end result. The line of questioning would be; "Infinite, the vitamins that would be most beneficial to _____________(name), are _______________(vitamins list), to what degree ____(%), and how often _______(**Time Factor Chart**)?"

Minerals & Elements Chart

Minerals & Elements: This chart is similar to the *Vitamin Chart* except that it has additional elements. Because of some of today's agricultural practices, our food can be largely stripped of much of its nutritive value. It is getting harder to acquire the necessary amount of vitamins, minerals, and trace elements from our foods for our bodies to assimilate for vitality and overall health. This is why we need to check on a regular basis to see whether we are getting enough from our foods and whether we need to supplement our diet with vitamins, minerals, and trace elements. This chart also contains some elements to check for excess. Excess can be equally unhealthy. The line of questioning would be: "Infinite, this body is in need of what mineral _________, to what degree ___(%)?", or, "This body is in excess of what mineral ________, to what degree ___(%)?"

Tissue Salts Chart

Tissue Salts: The Tissue Salts (cell salts) are the trace elements on the cellular level that are essential to the life and well-being of the human organism. Much research has shown that dis-ease and pain will result from an imbalance or deficiency of the tissue salts in the human body. This chart will help you to choose the needed tissue salt(s), and also aid in determining the potency level of the tissue salts needed for relief and/or balanced health. It is equally important to determine the appropriate timing (see Time Factor Chart) in this analysis for effective tissue salt use. The line of questioning would be: "Infinite, this body is in need of what tissue salt ____________, to what degree ___(%), at what potency ______(3X, 6X, 12X, 30X,.........), for how long ________(1 week., 2 week., 1 month., ...)?" Before going out and buying tissue salts we would suggest that you consult a naturopathic doctor, and read The 12 Tissue Salts by Esther Chapman.

Bath Chart

Bath Chart: This chart allows you to choose between the many different forms of therapeutic baths. In some cases, one or more of the listed therapeutic baths may be contraindicated for one's well-being. The questioning needs to indicate whether a particular type of bath is for the individual's highest and greatest benefit, and what degree of exposure would be beneficial. The line of questioning would be: "Infinite, for the greatest therapeutic effect, this body is in need of what type of bath ____________, to what degree ____(%), for how long ______(10, 20, 30 minutes ....), and best to repeat ___________ (1, 2, 3, times a day, week, month....)?"

Metals Chart

Metals: This chart can help to determine which of the metals would be beneficial to have on or around the body, and which of the metals may be creating a high level of metal toxicity within the body. This chart can also be used for prospecting for precious minerals. The line of questioning would be: "Infinite, which of the metals would be for the greatest benefit of this body ____________(name), to what degree ____(%)?", and, "This body has a metal toxicity of what metal __________, to what degree ____(%)?" For prospecting, the line of questioning would be: "Infinite, indicate which metals are present in this rock sample __________, and to what level of concentration ________(1/8, 1/6, 1/4, 1/2, 1, 2, ... ounces per ton)?"

Gemstones & Quartz Gemstones Chart

Gemstones & Quartz Gemstones: The use of this chart is a necessity for those who are wearing or using crystals and gemstones for therapeutic purposes. A great deal of the printed material on the subject of gemstones is vastly lacking in accuracy and validity. We suggest that you find out what is true *for you* when it comes to choosing crystals and gemstones for certain healing aspects and attributes. Each stone has a particular energy vibration which can affect the human body. Inappropriate stones at the wrong time can have a debilitating effect on the human body due to an overcharging of the nervous system. Because of all the misinformation circulating about the mineral kingdom, it is imperative that we learn how to use our Intuition to choose the appropriate crystals and gemstones. We need to know the appropriate gemstone for a particular need. It is equally important to find out the best place to wear a gemstone and how long it should be worn. The line of questioning would be: "Infinite, the best gemstone for this situation or application is ____________, to what degree ____(%)?", "The best gemstone for this dis-order, dysfunction, etc... would be __________, to what degree ____(%), placed on the body or worn for how long ______(10, 20, ...minutes, 1, 2, ...days, weeks, months), and the best place to wear this gemstone is _________(pendant(crystal termination up/down, fingers, wrist, solar plexus, pocket, etc...)?"

Blank Charts

Blank: There are blank charts so that you can make up your own charts to accommodate your particular belief systems, desires, or form of therapy. You will find that you could create charts on any subject to assist you in answering your most common questions. Regardless of your area of questioning (personal, business, interest, or hobby,) there are no limits to your questioning except your own self-imposed limits on your beliefs.

The charts and the lines of questioning will assist you in developing communications with the Intuition in a relatively short period of time. After you have worked with these charts for a while, you will find your left brain becoming very involved with the line of questioning and with the visual focus on the charts. After a few moments, you will find yourself slipping into a slightly altered state of consciousness in which your perception will feel defocused, while the questioning seems to continue with crystal clarity at a fast rate. At this point you may find some of the questions and answers to be different from what you may normally think about the situation at hand.

Sometimes, I have discovered more truths through the questioning process than from the answers themselves. In complete analysis of the physical, mental, emotional, and spiritual aspects of an individual, (sometimes taking as long as one to one and half hours), I have found myself in what feels like a time warp. There is no longer a sensation of time. When I finish the process, I am always surprised at how much time has passed. Typically the session leaves me feeling very energized, meditative, with a clear sense of having seen the truth behind the questioning. Most often the answers vary immensely from what I may have thought (intellect) was true at the the start of the questioning. There is usually the feeling of having seen the bigger picture, and how it all fits perfectly together for that person or situation, regardless of how it may appear. The choices that you, or the individual being tested, have made about certain life experiences become apparent. It is a matter of seeing clearly all of the options, and choosing accurately the one option that will be for your highest and greatest benefit, and will bring you on course with your life's purpose, desires, or goals.

With all tools of the Intuition, it is a good idea to keep a logbook with the results of the analysis, testing, measuring, or exercises. In this way, progress can be noted and the areas that need work will be revealed. These records can also help one to discover the area in which one is the most accurate or proficient. On a personal note, a record book will also reveal a great deal about your own personal growth as you practice your expanded abilities and beliefs.

Chapter 8

Distance & Map Dowsing

The Intuition is not limited to the physical plane and therefore is not restricted to Time and Space (Distance).

Map dowsing is a term applied to the locating of persons, animals, objects and substances either above, in, or below ground or water, from a distance, by means of a pointer, map, chart or photo, dowsing instrument, and the Intuition.

As discussed earlier, the advanced use of a pendulum from a scientific approach is called *radiesthesia* which means detecting, measuring, or locating the entire spectrum of radiations (energy). These can be objects, substances, minerals, plants, animals, or humans. If you are dowsing from a long distance away it is called *distant dowsing or teleradiesthesia..* Map and distant dowsing are forms of intuitive communication utilizing charts, diagrams, photos, documents, letters, signatures, or samples from the area of the quest. The application of distant dowsing can include searching for minerals, metals, precious stones, mines, tunnels, buried treasure, old monuments, foundations, graves, archaeological sites, artifacts, ships, schools of fish, lost aircraft, vehicles, missing people, bodies, pets, oil, water, money, valuables, and many other things. Map and distant dowsing are almost one and the same; the main difference is that one procedure uses a map, and the other uses a photo of a person, or a *witness* (sample). The "witness" consists of nothing more than a sample an object owned or worn by the person or a sample of the substance involved in the search. It is believed that the witness aids the dowser in focusing and concentrating on (or perhaps creating an affinity with) the person, object, or substance of the search.

The object, substance, or area of the search may be hundreds or thousands of miles away, or buried many feet below the surface. Some people may have a problem accepting this form of intuitive fact-finding. The thought of being able to pass a pendulum over a map and locate an object or person seems a little far-fetched to many people. However, when you first started to develop your intuitive abilities, the thought of being able to use the Intuition to *search* for some of the previous types of unknowns may have seemed almost impossible. Map and distant dowsing may seem inconceivable to some, but so do many other phases of the art of dowsing. As with most areas of the Intuition, there is a great deal that is yet to be explained.

Thought Forms

Where the thought goes, the energy flows.

Thoughts can travel great distance in basically no time, as indicated by Marcel Vogel's experiments with plants. Marcel presented clear evidence that our thoughts can have an effect upon plants from thousands of miles away. A summarized version of this experiment is as follows: Marcel had a laboratory full of plants hooked up to EEG (Electroencephlograph) equipment and strip-graph recorders. The object of the experiment was to see if his thoughts of love and light could be projected from thousands of miles away to these plants. He was scheduled to give a lecture in Prague, Czechoslovakia. His assistant was told to monitor the plants and that when Vogel got to Prague, he would determine what time of day he would send these thoughts of love and light to these plants. In other words, a double blind scientific method experiment was set up. When he got to Prague, some 10,000 miles away, he decided to send the thoughts of love and light to the plants exactly at 6:00 P.M. Prague time for five days straight. This is exactly what he did, though he did fall asleep one of the nights at the designated time and was unable to transmit the thought. Each time his intention was to send love from his heart and light from his brow center to his favorite plant in the laboratory 10,000 miles away.

When Vogel got back, he discovered that exactly 6:00 P.M. Prague time, his favorite plant had responded first, and subsequently the rest of the plants in the laboratory followed with their response. The plants had responded only four out of the five planned times. The response time of the plants was instantaneous. The great distance seemed to make little difference with the objective. The results of these experiments have had incredible impact on what we believed true about the properties and potential of human thought traveling great distance, and the response of plants to emotions. For extraordinary research on various realities within the plant kingdom, we suggest The Secret Life Of Plants, by Tompkins and Bird.

Science can not explain how thought and thought forms work, nor can it explain the rationale of mental activity that transcends time and space. There are only suggestions and clues from the scientific community. According to T. Edward Ross, during the *search,* the dowser will register brainwaves including beta, alpha, theta, and delta on an electro-encephalograph. All of these brain waves will register on the electroencephalogram at the same time. The dowser is therefore considered to be in resonance with the Schumann Resonance of 8 plus Hertz, which is the 'signature' of the earth. In other words, the act of dowsing reflects a full range of brainwave activity, and, the whole system with all of it's parts seems to be working together in an efficient functional balance which is consistent with the Earth's balanced energy.

Distant dowsing involves the ability to send one's Intuition to investigate a place, thing, or condition in question, and return to the sender with the desired information. It is the vital life force of the dowser that is able to transport itself to any part of the world or known universe with the incredible speed of thought. It moves free of the physical body and is able to intuitively communicate with the dowser through the movements of the pendulum.

Another perspective explaining how the Intuition is able to acquire unknown information in distant dowsing from maps, photographs and other objects, theorizes that rays are being emitted from the map, photo, object, or substances itself. It may be emanations from a photo or map that are actually a trapped light wave or ray. Since science cannot definitely explain how and why distant dowsing works, the potential distant dowser will need to come to a conclusion about his/her own belief system on the perspective of the physical, metaphysical, and divine aspects involved while dowsing from a distance.

If a part of your belief system is still having a hard time with all of this, think about your ability to attune yourself, or to come into sympathetic resonance, with something that is a few feet away. Now, expand that level of awareness so that you can see that it makes absolutely *no* difference on your intuitive or dowsing abilities whether an object is a few feet away or thousands of miles away. Our energy system allows us to broadcast, receive, and transmit information in the same way as a radio or television works. The mind, with *intention* and focused *concentration,* are the tuning devices like that of a radio receiver. The body is your antenna. By holding an image of the objective of your search, by formulating clear and specific intention, and by a clear and focused communication with your Intuition, you can connect with the object and access the information in response to your inquiry.

This intuitive ability is almost "reconnaissance" in nature. A part of our mind, (the intuitive part) combines with the universal life force energy that travels to the area of the search. Why does this happen? Every area, every object or person has its own unique qualities.

Everything vibrates at a specific frequency. This is why a *witness* is such value in dowsing. The *witness* acts as a vibration or frequency reference point held in the dowser's conscious mind. This, and the visual imagery of the map or photo, cause the Intuition to send the life force to travel to the location of the object or substance of the *search*, where it makes contact and returns with the information of the *search*.. The life force of the Intuition travels to the site of the *search* with the help of the map, photo, or sample by means of the directed *intention* of the conscious mind to the subconscious to find the matching frequency of the sample or witness of the same objects. As with all forms of dowsing, it is matter of establishing the sympathetic resonance of like frequencies. When contact at the *search* or *find point* is made by the awareness of the Intuition, there seems to be an almost instantaneous response.

One of the best ways to assure your conscious mind that map and distant dowsing works is to have an experienced map dowser do a water vein(s) test of your property from a hundred or more miles away with a map, a pendulum, and, maybe, a rock sample from the test area. Then at home, water dowse the area yourself. Most likely, you will come up with the same or similar results, proving to yourself that the unknowns can be realized without proximity to the test area.

The next step would be to do distant or map dowsing yourself for water or other materials on yours or someone else's land, so you can easily check for the accuracy of the results of your practice. As long as you can maintain a clear, concise, simple, and accurate image or visualization of the object of the search, you do have the ability to send a part of your mind hundreds or thousands of miles away in a matter of seconds to the object or person of the test, and return with accurate answers to inquiry. This does take a certain amount of practice; however, by using the *Pendulum Charts* and a focused, determined, and trusting mind, this task can be greatly simplified and accomplished in a relatively short period of time.

One of the most fascinating long distance or map dowsing cases involves a man by the name of Verne Cameron. In 1959, Verne

Cameron contacted the United States Navy and told them that by using a map and a pendulum, he would be able to locate the entire submarine fleet. In March 1959, Vice Admiral Maurice Curtis wrote a letter to Mr. Cameron: *"I am advised you believe you may be able to tell the location of all the submarines in the world's water by a technique called "Map Dowsing". It is suggested you be given an opportunity to confirm your ability on the subject"*. Soon after the letter, Cameron met with the Vice Admiral in Southern California with the rest of the Navy brass to demonstrate the technique.   Within a few minutes Cameron located every single submarine to the total amazement of those present. He also located every Russian submarine around the world. Despite the success of this test, Cameron did not hear from the Navy or United States Government until years later under different circumstances.   Cameron had been invited by the South African government to their country to discover various natural resources using his Intuition and a pendulum. When he applied for a passport, to his great surprise he was turned down. He investigated the reason for the denial and found out that the Navy had contacted the C.I.A. about his submarine dowsing ability. The intelligence agency labelled him as a security risk and did not allow him the freedom to travel outside the country for fear he might reveal top-secret military information.   But even the military evolves. During the Viet Nam conflict, McNamara, then Secretary of Defense, used a pendulist to locate underground tunnels, landmines, and ammunition dumps. Today, we hope that the Intuition be used only in ways that will be much more beneficial to all.

Map and distant dowsing are easy methods for city and suburban dwellers to use as practice for developing their intuitive abilities. It is possible to sit in the comfort of your own home or apartment and dowse for far-away places, or even a vacant lot down the street. This method can be very helpful in finding the perfect house or piece of land to buy for you and your family. I have used this method for every home I have lived in the last 15 years. I first find the perfect time for the move, and then find the perfect location (specific area). I have always been amazed at how accurate this can be: down to the specific weekend determined some 2 to 6 months in advance. This technique has allowed me to be first to show up at the most perfect house and has resulted in me getting it. My friends are always wondering how I get such great houses, perfect settings, ideal living conditions, and at a cost much lower than others of equal value.

Distant Dowsing Preparation

The ideal condition for map and long distance dowsing is usually early morning when your energy level is highest, and when there is usually the least amount of noise and distractions. Do it only when you feel bright and energized. If it is a gloomy day and you are the type that can be drained by that type of weather, then wait for blue sky and sunshine, if that is what it takes. Some people find that late night is appropriate for them. This is because of the great decrease in environmental noise and air-wave activity from radio, television, and microwave transmission.

When dowsing with a map, you can use a simple hand-drawn map as well as purchased ones. Most black and white or colored maps, charts, diagrams, or photos will prove satisfactory. However, for accurate dowsing, remember that maps not drawn to scale can throw off the find by a distance of several feet to several miles. If your map is scaled too small, you will have difficulty in pin-pointing precise areas. For this reason, we prefer a blown-up version of the concentrated area of the search, accompanied by a small scale map covering the surrounding region.

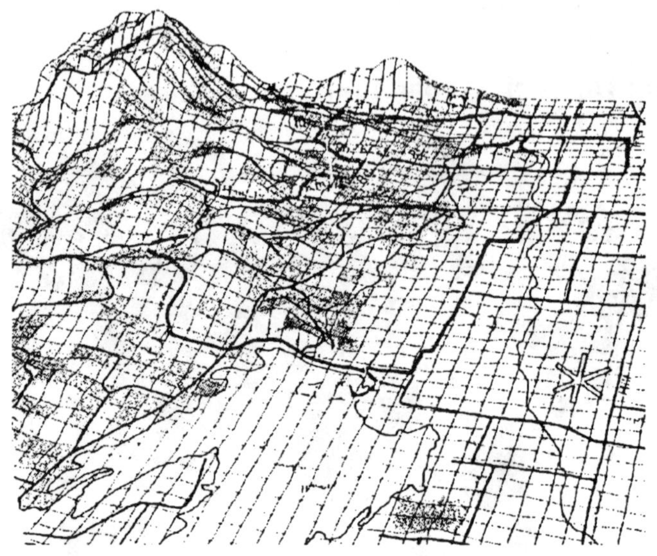

Topographical map with quadrant grid pattern.

In map dowsing, you may be surprised to find your pointer directing itself off the edge of the map. In this case, the location sought may be found on an adjacent map, or by adding a blank sheet of paper. For instance, the course of an underground stream may easily be projected on the adjacent map and it may be necessary to hand-draw on a piece of paper the remaining section of stream. The map or papers are secondary. More important by far, are the quality and clarity of questions and the answers given by the instrument and pointer.

Often, you may be dowsing a map that was previously dowsed by another person with similar *search* objectives. This is a very important issue that can influence negative effects upon the results of your dowsing findings. You will need to find a form that works for you in ruling out, neutralizing, or protecting yourself from other existing thought forms. We have already shared numerous methods of surrounding oneself and the object of the search with protective *light* through breath, affirmations, prayer and focused intention. Find a method that works for you so that you are not influenced by extraneous variables that may inhibit your own effectiveness.

Preparation: It is important to prepare yourself mentally for map dowsing, and to be specific in what you are looking for. BE ALERT for clues as you proceed, and your Intuition will be guiding the instrument in your hand and prompting you. Be prepared to record and evaluate your findings with utmost thoroughness.

Start by tuning in or getting a feel for both that which you wish to find and the general area under consideration. Orient the map to true North-South. With a ruler, draw lines radiating out from a single point, subdividing the search area into smaller segments like a sunburst pattern. The single point that you choose as a starting point should be a point on the map that can be easily found, such as a landmark, river, surveyor marker, road marker, etc. Focus your concentration on the objective of the search while looking at the map of the general area. Avoid thinking that deprives you of concentration. Free your mind of any pre-conceived ideas concerning the pendulum's movement where you might expect to find the objects of the search. Now, defocus your eyes from the map by a means of looking past or through the map using your peripheral vision to register the results of each test. This will help to keep your eyes from catching on one particular area for too long. Noticing one spot in particular may cause you to lose your concentration and possibly give you false results. Through your peripheral vision, watch for any exaggerated or excited action of the pendulum as the pointer moves over "live" areas on the map, photo, or diagram. Feel for these spots and remember, keep your thought continually on the quest of the person, place, or object sought. Be alert to any change in direction or swing of the pendulum as you proceed. With a wooden pointer, or the index finger of your left hand, point to each area.

Now begin your line of questioning:

- "Now is the perfect time and place to be doing this search?"

- "It is within my ability to Find _____________ (object of test)?"

- "It is to my highest and greatest benefit to find ___________ (name, substance, etc.)?"

- "I can find _________ (name, or substance) in this segment_____ (Yes/No)?

- "The probability of Finding __________ (name, substance, etc., is what degree in this segment ______ (%)?"

You continue with this process until you have narrowed it down to a smaller area. Then we would suggest you get a quadrant map of that particular area and narrow it down again. From there you work from the point of your radiating lines; make sure your map is oriented on true North, and begin working with your questions: "From this point, the object that I am seeking is what direction or degrees from known point_____?", then draw a line from the point on the map in the direction or precise degree. Then ask: "I will find the object_______ within how many feet, yards, miles, from this point _____?", now draw the intersecting line on your degree line. The *find point* is where the lines intersect. Now go and recheck your *find point* and see if results are accurate.

The map dowser must remain somewhat passive after adequate preparation has been made to avoid a forced or contrived situation. One's ego can nullify map dowsing success. It is better to map dowse in a surrendered, objective state of mind, employing the Intuition with confidence. Avoid the urge to find correct answers without first giving your Intuition ample time to investigate and evaluate the situation. Some of the greatest locating failures have been caused by not realizing that a few moments of concentrated thought imagery backed with desire (wishful thinking) can make a totally false location look like a genuine find. We highly recommend that you refrain from making up your mind about the results as you would like them to be.

Unfortunately, you can create a positive thought picture so vivid in your mind that the person, object, or substance seems to be there. The active subconscious mind can at times work with a very subtle level of delusion. This is not done with malice by any means. However, the subconscious mind is that part of the mind without reasoning ability; it does wish to please and can exaggerate to do so. It is best not to concentrate on *where* a certain person, object or substance must be. By doing so, you can create a false thought image and transpose its reaction through your instrument. The dowsing device in your hand cannot differentiate; it can follow this influence and be led to a false site. This problem has, on more than one occasion, shaken a individuals' confidence in map dowsing.

As always, before investing a great deal of time and money into planning around a *find point,* do go and check it out. We just want to remind you about wishful thinking and visions of grandeur. Use as many ways as possible for checking out a particular *find point.* Sometimes the Intuition will take you to a *find point* to show you something else that you need to know, or something you may have overlooked in your preliminary test. Sometimes, but rarely, **"X"** marks the spot. Using a combination of intuitive tools, such as subtle sensing, pendulum, and dowsing rods, can be helpful as a way of verifying your test results. In other cases, it would also be advisable (before digging some 30 feet in attempt to find a buried treasure chest), to bring in other tools such as a quality metal detector, or other means of verification.

If you were to continue your questioning once you located the *find point,* you could ask: "I will find this object ___________ located ___________ (how many inches, feet, etc...) below the surface _________?" Now you may have found your treasure, though I would like to caution you on treasure hunts. These methods can work on treasure hunts, and, you can be successful with it. However, I have seen some dowsers taken down the yellow brick road of illusion because of emotions invested in the outcome. To use the Intuition and the pendulum with accuracy, one does need to be careful about appropriately using their abilities for capital gain.

Problem Solving—House

Distant dowsing can be an incredible tool for intuitive problem-solving. An interesting application of a long distance dowsing happened some years ago when I had my construction business. I was working on a complete remodelling of a house, and for some inexplicable reason, the customer lost electricity to half of the kitchen. The electrical sub-contractor spent two days tracing down every possibility of the cause and found nothing. Another electrician was called in who also found it to be a mystery. They were ready to wire in a whole new line and bypass the old one. Meanwhile, I took Polaroid photos of the house; I went home that night and spent about a one-half hour doing a long distance photo test. I determined that on the front door of the house by the left side of the kitchen patio door, in the second panel to the left and the sixth nail over we would find the cause. The next day, I discreetly told my customer and the sub-contractor what I found, and of course they looked at me as if I were crazy. I could tell they were not going to act on the information I had given them. So I said that I would pull the wood paneling off myself, on my time and expense, and have a look. Meanwhile, I was thinking, "I sure hope I'm right, because this could look pretty bad if I'm not!" What I discovered when I pulled off the wood paneling was that, sure enough, the sixth nail over had severed one side of the electrical line. Everyone thought it was great that the problem was solved, but even to this day I bet they think the procedure in finding the solution was a little crazy and they did not really believe it anyway. I quickly played the story down, for it was definitely beyond their comprehension.

There are three messages in this story for you. First, there was a definite need. Secondly, we do know how to determine the unknown from a photo over long distance, and third, these techniques work best when you keep the ego out and do not try to impress anyone with your abilities.

The following is a form of distant dowsing that you can apply to your own house, a house you plan on buying or renting, for yourself or others. A photo or plans of the house will do just fine as a witness.

House Inventory:

Dowsing Report on: ____

Date_______________________

A. Roof:
Shingles/Slates: do they need replacing in next________ years.
If damaged, where? _______________________________________
Flashing: _______________________________________any leaks
potential problems____
If so,where ____
Gutters: adequate ____
problems _________________where ________________________
Downspouts: problems _________where __________________
Rafters & Braces: any problems ____________________________
where ____
Soffits: good repair & adequate___________________________
if not, where ____
Ventilation: is this adequate under roof ___________________

B. Insulation:
Ceiling: is it adequate ___________________________________
if not, above which rooms ________________________________
Walls: is it adequate ___________________________________
if not, where ____
Basement: is it adequate_________________________________
If not, where ____

C. Exterior Walls:
Brick: are there crack __________________________________
where ____
Is pointing required ___________ where _______________
leaning or bulging____________________
Stucco: does it need repair _________ where ______________
Is it solidly affixed____________________
Siding: needs repair ___need painting/stain _______________
Caulking: need around windows __________________________
around doors ____

D. Entrances:
Doors: is insulation O.K. ________________________________
are they hung properly___________________________________
are locks secur ____
are there screens _______________________________________

Windows: are frames O.K. ___________________________________
     do they fit O.K. _______________________________________
     do they need painting__________________________________
     are they burglar proof_________________________________
     need of caulk ____
     are there screens _____________________________________

E. Heating:
     Furnace: ___________________ does it need replacing in next
years
     Piping: if hot water, are there leaks_______________________
     is there corrosion _____________________________________
     If hot air, is system operating ___________________________
     O.K. ____
     is humidity system O.K. ________________________________
     is dehumidifying system,
       or air purification system working properly _________
     Fireplace/Woodstove: are they in good order _____________
     does internal masonry in chimney require repair _________
     does external chimney require repair _____________________
     is there a good draught_________________________________
     any smoke-leaks out into room or under roof_____________

F. Basement:
     any structural defects __________________________________
     any damp spots ____________________where ____________
     adequate height ______________water leaks____________

G. Sewers:
     Septic: problems inside house____________________________
     outside ____
     Storm: problems inside (such as backing up) _____________
     problems outside, perimeter drains ______________________
     storm drain outflow____________________________________

H. Electrical:
     Service: properly fixed to house_________________________
     adequate rating for house (amps)________________________
     Fusing: is house fitted with adequate fuses or circuit breakers
     Wiring: correct type/rating_____________________________
     Fixtures: are lighting fixtures and wall outlets safe_________
     if not, where ____

I. Gas/Oil:
     if any, are these supplies in good working order _________

J. Grounds:
     Garden: are there fencing problems ____________________

      where ____
      any unsafe trees/shrubs ______where _________________
      toxic substances in grounds ______________________________
      where ____
      major pest problems ______________________________
    Driveways: ______________________________________
      is surfacing adequate ___________ problems___________
    Garage: sound in structure___________doors fit___________
      roof O.K. _______________________________

K. General:
      Is area liable to flooding __________earthquakes ________
      possible zoning changes in next 5/10 years ____________
      possible development of area in next 5/10 years ________
      possible impact of development ______________________
      of new highways in next 5/10 years ________________
      Is property close to High Voltage transmission lines _____
      are there any "noxious rays"
        running under or through house __________________
      caused by "streams" ______electro-magnetic___________
      transmission ________________ ore bodies __________
SHOULD YOU MAKE THIS INVESTMENT_________________
In order to undertake this survey, all that is required is a floor plan of the house, or a photo of house. Set the plans and/or photos and map of the area in alignment with compass directions. Begin to access a 'feel' of the house and location. Visualize it in your mind with as much detail as possible. When you have felt or seen the entire picture, begin your house inventory.

Dowsing Photographs

You can apply your intuitive or dowsing abilities to a photo of person(s), places(s), objects(s), or substances(s), living, dead, or inanimate as though they were in your immediate presence. You can determine the sex of the individual, or know whether he or she is living at the moment the picture is dowsed. Even if the photo is obscured, the present location of the person as he/she moves from one place to another can be determined.

All that is needed to do distant photo dowsing is a pendulum, *Pendulum Charts*...if you have them, a photo of the person, place or object of the *search*, and your faith and determination that this is within your abilities. When using a photo in testing a person, make sure that there is no one else or any animal in the photo with the person being tested. It does not matter how old the photo is; it's only purpose is to help you to tune into the frequency of that person, place, or animal.

Before you start, **always** surround yourself and the person of the inquiry with the Clear-White Light and Love, and ask, for example, "Infinite spirit or Infinite Intelligence, allow me now to be tuned into this person. It is to this person's highest and greatest benefit for me to do this test?" "It is of this person's will for me to do this test and I have their permission?" "Let this be of thy will and not my will." Only continue if the pendulum answers are a resounding *yes*. If the pendulum answers to any of this line of inquiry are anything but a strong Yes or clockwise gyration **DO NOT** continue on. As with everything in life there is a fine line between balance and imbalance, light and dark, the good and the not good. You <u>do not</u> want to cross over no matter how tempted you may be, for the outcome would not be for either your, or the other persons, highest benefit. It could potentially only hurt you in the long run. With a *yes* answer, you can continue on checking all the various aspects and attributes of the physical, mental, emotional and ethereal bodies. In working in this realm, I would like to assure you that we really do have access to this infinite level of knowledge. As with any skill, the more you work with it the more proficient you will become. Your goal is having full access to an effective and accurate Intuition. Once again, we would like to remind you that most of this skill has to do with intention, concentration, staying focused, questioning appropriately, and believing in your infinite abilities.

Dowsing Photographs: First place your left hand above the photo. Slowly raise and lower your palm; meanwhile 'sense' the energy focal point (the highest level of intensity) above the picture. Mentally pose one question at a time: "Is the subject in the photo now alive? Traveling? Married? etc., swing your pendulum and await the *Yes, No, Maybe*, or more-or-less response.

Locating Loved Ones

You can use this technique to locate loved ones. First of all, it helps to know a little information about the person being sought. It also helps to define exactly whom you are seeking. A name may not necessarily be enough. If the parents are asking for the information, then state your request like the following: I am seeking the physical location of the individual known as Mary Jane Smith, born May 23, 1969, who was living at ....address, the daughter of John and Sue Smith, 244 Main St. Podunk, Iowa. "Infinite, may I locate this particular individual?","Can I locate this particular individual?","Should I locate this particular individual?", "Am I now ready?" Proceed only if you receive *yes* answers.

Your job will be easier if you have several maps and some yardsticks available. When looking for the location of a person for the first time without any clues, it is best to start with a world map. Begin the search with a world map, then continue with a regional map, and then finish with a local map with the top of the map orientated to true North. Then ask "Infinite, please indicate, on this map, the location from this point on the map, where the individual I am searching for is located at this point in time." The pendulum will swing to point in a given direction across the map. When the motion has stabilized in one particular direction, gently lay a yardstick down in that precise direction. Go to another side and ask the same question. Then again with the third side, and finally the fourth side with the same question. The intersection of all the yardsticks will give you an exact location. Place a mark on the map at the spot. Then switch to a regional map of the indicated area and repeat the process. You can continue to use more detailed maps until you can literally locate the individual in a particular building on a particular street! It helps to write in the date and time by each of the marks on each of the maps. Dowsing daily or weekly will add confidence to your dowsing skill and also will indicate the path of the individual. Using the marked map, you will be able to check the accuracy of your dowsing at a later time.

Also consider that in some cases knowledge can be a terrible burden. Do not ask for information that you would rather not know or that you do not need to know! People can be hurt just as much by too much information as too little!

Missing Person Search

Never volunteer! Always wait to be asked. If this is something that you would like to be of service with, then we suggest that you work with missing persons you hear about, follow the news, go through the steps, and watch the outcomes.

How do you establish acceptance? Set up search situations to test your ability and when you begin to see some relative and consistent accuracy, contact the appropriate Search and Rescue groups and explain to them how you will attempt to assist them.

When you have been called in to participate in a search and rescue, prepare yourself for some apprehension from those in the group. Start your work by collecting as much information as you can about the missing party. If a picture will aid you, get it. Where was the person last seen, by whom, when, what were the intentions (if known) of the individual at that time, what are the habits as they pertain to the disappearance? Feed all this into your mind-computer so that you have a sense of the missing person as much as possible

Seclude yourself if you work best alone, or have a second person to help you with the first steps of Map Dowsing. Get detailed maps of the area. If you are looking in a city, get a block-by-block map; if the country, a geodetic topographic type; if the person has traveled you may need state or regional maps. As for the actual procedure, use the previous methodology Locating Loved Ones, or earlier in the book, Direction Finding.

Of all of the dowsing methods, distant and map dowsing may be the most difficult to comprehend and believe in. Whatever applications you intend to use these techniques for, practice will enhance your confidence and your ability. Your preconceptions and beliefs may be deeply rooted, so be patient.

As with all types of dowsing, work to neutralize your ego involvement. Humility can go a long way to assisting your practice and results. It is always important to ask permission to dowse and to surround yourself and the person or object of the search in the White Light and love (see Chapter 4). It is especially important in distant and map dowsing so that you not interfere with the desires, lessons, or free will of the person for whom you are searching.

Chapter 9

Success, Why Not

All dowsers deal with errors, mistakes and setbacks at various points in their dowsing development. However, the frequency of error reduces with time as the dowser gains experience, and as he/she becomes clearer from mental, emotional and physical interference. An understanding of the reasons for dowsing error can help veterans, as well as amateurs, to avoid repeating the same mistakes. Often we tend to focus on successful experiences rather than finding the causation of our errors. In this assessment, the influences of these dowsing mishaps or failures are being labeled as *'jeopardy influences'* that have been divided into levels associated with the various aspects of our consciousness. *Jeopardy influences* are potential problems or stumbling blocks that you may run into on your journey into proficient dowsing.

Jeopardy influences are internal or external conditions that can become obstacles to effective results. A protective light energy field may keep one relatively clear in spite of the presence of several of these *jeopardy influences*. A dowser may even have enough energy to override the conflicting influence of certain *jeopardy influences*. Yet it is possible that the existence of even one *jeopardy influence* could cause some degree of interference and affect accuracy. The point is that it is important to be aware of these *jeopardy influences*.

By being consciously aware of them, the dowser can work to control, clear, or neutralize as much as possible. *Jeopardy influences* are of various types and may be grouped according to physical, mental, emotional, and psychic/spiritual sources.

Jeopardy Influence Dowsing Inventory

Physical Internal Influences:

- ☐ Hunger and/or fatigue

- ☐ Improper diet, poor nutrition, or gluttony

- ☐ Vitamin and mineral deficiency

- ☐ Infection, illness, disease, and trauma

- ☐ Chemical/pH/hormonal/glandular imbalance

- ☐ Drug use of all kinds, including caffeine, tobacco, alcohol, and medications

Physical External Influence:

- ☐ Electrical or electromagnetic fields which are disharmonious

- ☐ Audio and/or visual distractions or disturbances

- ☐ Noxious negative earth energies

- ☐ Weather disturbances

- ☐ Threats to physical safety, insecurity, stress

- ☐ Dirty, smelly, or otherwise offensive surroundings or environment

Mental Internal Influence:

- ☐ Depression, anxiety, and stress

- ☐ Addictions, obsessions, compulsions, and perversions

- ☐ Projection, denial, and other defense mechanisms

- ☐ Intellectual conceit, egotism, and vanity

- ☐ Disbelief, doubt, or not being open to (fear of) higher levels of consciousness

- ☐ Desire to control, reactionary, self-pity, self-doubt, selfish motive

- ☐ Negative thought patterns including judgment, greed, self-centeredness, and others

- ☐ Biased interpretation of response according to limited belief system

- ☐ Lack of focus or concentration

Mental External Influences:

- ☐ Subliminal messages

- ☐ Hypnotic domination

- ☐ Violent movies/films, books/magazines

- ☐ Psionic devices

Emotional Internal Influences:

- ☐ Emotional attachment to the answer, i.e. strong aversion or wishful thinking

- ☐ Limitations regarding how much information you can handle at the time

- ☐ Undisciplined emotions: fear, anger, doubt, jealously, guilt, hate, rage, envy, hostility, resentment, dissension, unforgiveness, worry...

- ☐ Emotionally painful or embarrassing information which the subconscious mind may withhold or exaggerate

- ☐ Traumatic memories (especially sexual abuse/molestation) which may cause the subconscious to cover up

- ☐ Fear of or resistance to the answer

Emotional External Influences:

- ☐ Abuse of all kinds

- ☐ Experience of injustice, violation, abandonment

- ☐ Withholding love

Psychic/Spiritual Influences:

- ☐ Unbalanced chakras

External Influences:

- ☐ Presence of skeptics

- ☐ Entities in the auric field

- ☐ Locations such as graveyards

- ☐ Sites of traumatic incidents, such as murder or rape

- ☐ Psychic attack, either by conscious or unconscious

- ☐ Season or time of year, time of lunar cycle, negative astrological influence

☐ Empathetic or sympathetic pains (picking up another's suffering)

☐ Psychic intrusion by disharmonious forces: human earthbound disincarnates, non-human entities, fallen angels, devils, satanics and demonics, elementals, thought-forms, evil adepts and masters of sorcery, jealous and manipulative "gurus," and occasionally, animals.

These potential interferences are, of course, by no means inclusive, but can assist you to sort through possible influences in your process of development. Physiological factors are more easily dowsed for by yourself. Mental and emotional influences, on the other hand, are more difficult, and are better dowsed for by another. This most often is because it is difficult to be objective to the answer or outcome. Psychic and spiritual influences are by far the greatest problem to recognize and may also require the assistance of another dowser to determine the existence of such disharmonious presence.

Solutions

Of course, we would not just give you the problem without offering workable solutions. There are many ways for clearing out subconscious blocks, traumatic imprints, possessions, or psychic intrusions. The forms of therapy associated with such dis-order and dysfunction are beyond the scope of this assessment. Nonetheless, there are some forms of transformational therapy that we have seen to be most effective in restoration of one's total well-being. One such form of therapy is the Spiritual Response Therapy developed by Robert Detzler in <u>Your Mind Net</u> [1]. This approach involves naming a specific problem area and dowsing to determine the number of reasons contributing to the dysfunction. The following list of categories, or "keys", are to assist you in determining the reasons, purpose, or awareness for various conscious or subconscious negative programs.

The Keys:

- ☐ Identification

- ☐ Imprint

- ☐ Experience/Trauma

- ☐ Benefit

- ☐ Conflict

- ☐ Self-Punishment

- ☐ Organ Language

- ☐ Past Lives

- ☐ Future Lives

- ☐ Disincarnates/Entities/Thought Forms

Once identified, the negative conditioning or imprints can be cleared by the appropriate affirmations, prayer, or therapy. However, each conditioning or imprint cleared or neutralized must immediately be substituted with a positive program, condition, or energy. You must literally fill that space which the negative conditioning occupied. (Affirmations are particularly good for this.) Review chapter 4 and 5 for some ways to remedy negative conditioning. Your pendulum and pendulum charts will also assist you in determining therapy modalities to use, and an appropriate time to use them.

Negativity of any kind will cause pollution within our being, in our environment and also within our bodies. Awareness and protection are the keys to success against *jeopardy influences*,. Denial, blame, or martyrdom will not bring success on any level in overcoming these stumbling blocks.

No matter how detrimental a possession or entity may seem, we must always ask permission first before releasing disincarnates. "Some people have accepted disincarnates as a part of the learning and overcoming program they established for themselves to challenge them in this life, and (we) have no right to interfere" [2]. In some circumstances, the disincarnates may have set up a situation, with the permission of the host, in order to learn certain lessons without having to incarnate. Even in the case of non-human entities, one must be aware that "some people want to keep the possession in order to be special or powerful" [3].

Untruth can also come from the subconscious mind. Although it can appear deceiving, the purpose for untruth will always be for one's higher good. This may happen in the interest of your protection if you seek information which you are not mentally or emotionally prepared to hear. For example, your subconscious may try to cover up particular traumatic memories, such as sexual abuse or molestation. However, questions asked at the proper time will most likely to receive a true response.

Bogus answers may result from asking out of wrong or selfish motives, or imploring trivial information. Proper motivation means that the dowser is coming from a place of true need, is sincere, and has the highest good of all concerned at heart. Furthermore, the Intuition may give distorted information when it detects mistrust, skepticism, testing, or lack of respect on the part of the dowser. The appropriate attitude is crucial, even in the observer, before the dowser can receive reliable results.

Accurate information about the future, as previously discussed, is the most difficult to obtain from the Intuition. The ego will sometimes manipulate the future once it is known, against its own best interest. This may be the result of some subconscious self-sabotage or self-punishment programs or conditioning. When permission is granted to inquire about the future, Robert Detzler advises that it is best to word questions in terms of "based on the divine plan for one's life." [4]

Having some awareness and knowledge of the nature of spiritual guides can be most helpful in forming realistic conception of them.

Guides can come from many different dimensions, realities, forms, and planes of existence. Their expertise is equally as varied as is their level of proficiency in connecting with humans. They also can have limitations and be at different stages of their own evolutions. What is important, and this is perfectly clear, is that they are of the Light. Guides at higher levels of awareness can be recognized by their attempts to help you empower yourself. If they are truly your guides, they will assist you in remembering who you are, to let go of fear, and to love yourself and others. Highly evolved guides will encourage your independence and self-sufficiency. They will assist your decisions, but they will not make them for you. Rather, they will point you in the appropriate direction. Spiritually evolved guides will encourage you to trust your own Intuition. Rarely will guides predict future events, for they are careful not to take away your lessons. Their messages are positive and uplifting, expressing truths in precise, and almost simple, terms. Yet, they are humble and will admit their word is not absolute.

The lower level guides, by contrast, are pretentious and talk in profound sounding trivialities. Most often they lack greater understanding and genuine wisdom. Usually they are attracted by one's curiosity. They like to stir up emotions by predicting disastrous events, or to flatter people's egos with grandiose predictions. Low level guides like to entice one into giving power to them and one is usually left feeling fearful or depressed. Some show outright offensive misbehavior with no commitment at all to spiritual growth or to one's highest and greatest benefit. Keeping as free as possible of *jeopardy influences* is the best insurance for not attracting them in the first place.

If there is the least question in your mind of interfering *jeopardy influences*, always check to see if additional protection is needed. Even after all *jeopardy influences* are eliminated, there is still the possibility of error due to improper procedure.

Errors Due to Procedure

- ☐ Lack of focus or concentration

- ☐ Not checking to see if clearing, balancing, or protection is needed

- ☐ Failing to determine the identity of the source of the information and from where it comes

- ☐ Neglecting to ask the questions Can I?, May I?, Should I?, Am I Ready?

- ☐ Failing to check for interferences arising during the search

- ☐ Doubt, lack of trust or respect for the source, testing the source

- ☐ Incorrect motivation, lack of true need

- ☐ Lack of specificity in asking the question, incorrect wording

- ☐ Not checking for emotional or mental interference with the answers

- ☐ Incorrect interpretation of response

In summary, the preparation of the dowser begins with the proper attitude of faith, trust, respect, and a sincere motivation of true need. The most common error is the lack of being specific in asking the question, or incorrect wording of the question. Usually this error has to do with not being specific enough about time and place, or about the language with your pendulum. Keep in mind that the subconscious mind takes everything totally literally. One should take care to define to the subconscious which reality you are directing your line of questioning to: the actual physical reality or that which is imagined, potential, alternate, or subjective.

Finally, learn how to protect yourself from mental, emotional, and physical interference. Asking if you can does not always mean that you will avoid interference, or that some precautionary steps are not needed before you proceed. Lower thoughts and negative emotions must be kept to a minimum and transformed toward the higher mental and emotional realms.

By doing your work to enhance your personal growth, you will be paving the way for clearing out negativity. Reestablishing positive patterns where the negativity once was will assist you in your personal growth, will help to minimize *jeopardy influences*, and increase quality of living.

Errors and mistakes may still occur, but we can make them our friends...our teachers. We learn more from our mistakes than we do from our successes.

---

[1] Detzler, Robert E., 1988, <u>Your Mind Net</u>
[2] Detzler, Robert E., 1988: p242, <u>Your Mind Net</u>
[3] Greene, Linda L., 1990: p64, <u>Learning the Secret Language</u>
[4] Detzler, Robert E., 1990: p2, <u>Your Mind Net</u>

Conclusion

Throughout this book, I have shared with you many techniques to allow you easy access to your Intuition. With patience and practice, I believe you can successfully tap into your inherent dowsing skills. Your success will be dependent on your pendulum language, your beliefs, (whether expansive or limited) your technique, and jeopardy influences (whether external or internal.)

Dowsing is like many other activities; you learn more from your mistakes than you do from your successes. Each one of those mistakes must be understood as a positive step forward in the refinement of your dowsing skill. It is important that you make an effort to understand *why* any particular mistake occurred. What were you thinking at the time? Had your mind strayed from its image of the search? Had you made any incorrect assumptions? Realize that even though you may have made a mistake, you did receive a response to something. Your goal is to determine what that "something" was and why it was an error. The problem is not one of whether you can accurately dowse or not; it is one of refinement and clarity of the questions you formulate and the image you hold as you go through the physical dowsing process. All dowsers deal with erroneous results at various points in their dowsing careers. However, the frequency of error reduces with time as the dowser gains experience and becomes a clearer channel. Having a level of 80% or greater accuracy is definitely an accomplishment. In other words, it is okay to be less than 100% accurate with the results of your dowsing work.

Proof is sought as a mark of progress, and a track record of successes AND failures illuminates one's expansion into dowsing's many domains. Keep a logbook with the results of your exercises and experiments. In this way, progress can be noted and the areas that need work will be revealed. This record can also help you to discover the area in which you are the most accurate or proficient. On a personal note, a record book will also reveal a great deal about your own personal growth through your expanded abilities and beliefs.

Utilizing the Intuition with the intention of helping another person is much easier than one might think. The main ingredient is one's intention. From the proper perspective, an open heart and a loving, caring intention to help others, (with their permission), you can tune into another person anywhere in the world. Keeping yourself focused from your heart and brow center, with your intention always for the highest and greatest benefit of others, gives you a protective barrier of love and light that will allow you free access to the Infinite.

Dowsing is both art and discipline, and as such, you will benefit yourself and others by being as clear a channel as possible. This is where the discipline comes in. Daily personal growth work to identify, clear, and neutralize negative patterns, replacing those negative patterns with positive nurturing patterns, can only impact your life in an uplifting way—assisting you in your day-to-day living, and also in your skills to externalize your Intuition.

Becoming involved in your personal growth work requires pranayama, or control of breath. This is the one thing that will assist you the most in your development. Your level of consciousness is directly related to your breath and how you breathe. It is the attention to breathing that is important. Deep meaningful breath, which has to be balanced both ways, in and out, is a necessity. Have you ever experienced some days when, so to speak, you were sitting there all day holding your breath? You forgot to breathe in/out, didn't you? When you get to the end of one of those days, you're usually exhausted from lack of oxygen and prana. Some of us hardly breathe at all. We breathe in, or we get caught in the world of fear or attraction and our breath almost stops totally. We have to remember every day, to take a little bit of time to breathe consciously. This is called self-remembering, and it is part of becoming more conscious of yourself, of the planet, and everything else. Stop and just take one breath and look and see what you haven't seen before in the space in which you are standing. Your reality in this world is strictly your impressions, and breathing consciously will open your eyes and your mind. Learning to breathe is a responsibility—a mandate, and is essential, if you are to work at these higher levels of consciousness.

We all have the ability to heal ourselves. The major difference between one person and another is limited or expansive belief systems. Some people hold more expansive belief systems, and because of this, they have higher levels of potential energy—they are able to transform the ever-present subtle energies surrounding them into healing energy for themselves or others. This transforming power is directly related to our breath— by breathing consciously we can enhance this energy system within our bodies.

Concentrating on our breathing—consciously raising our energy levels and our personal awareness—are the initial steps toward healing ourselves or others. (This is "remembering" our innate healing abilities.) This is the first step in changing consciousness. Changing consciousness is like tossing a pebble into still water; the pebble drops, then concentric circles emanate out from the center. This is the effect that raising our consciousness will have on the world around us.

The responsibility to change consciousness lies with each individual. How do we change consciousness? Begin by eliminating negative beliefs. Take responsibility for your own life. Release old patterns, vicious circles, imprints, and debilitating conditioning. Replace those old patterns with new, positive, and nurturing patterns. Raise your energy level by conscious breathing, meditation, affirmations, and visualizations.

Consciousness begins with each of us. Consciousness is awareness, and internal perception. The state of consciousness from which we view the world and interact with it is crucial. The state of our consciousness determines the world as we know it, both what we perceive the world to be, and what we project to contribute to that world view.

Reshad Feild says that we are transformers of subtle energy. But realizing that there are many stages of development, we want to know what kind of transfomers we are. How clear are our channels? How well have we polished the mirrors of our hearts? If we as dowsers wish to be of service, at what level do we understand service? How close is our service to reflecting Divine Love? Those great beings who manifest that stage communicate health and life by their presence alone. We are most fortunate, for we have been made aware, through learning to dowse, of the divine spark within everyone. We have been made aware of our potential to move beyond what we had accepted as the limitations of our world in a progressive expansion of consciousness which is really only limited by the limits we put upon it. *"The only limitation my mind has is belief in its limitation,"* said Dr. Abdui Aziz.

Dowsing is about seeing at a distance—beyond the five physical senses. As we climb a mountain, we get a wider view. And so it is with the expansion of our consciousness beyond that of our own personal, limited egos. Becoming a better dowser means learning better how to live in harmony with the earth and all she supports. This is developing planetary consciousness. It means expansion through learning the divine intention and then helping to manifest it as transformers of subtle energy. Discovering the divine intention and helping it to manifest, that is what service and helping really are about.

By helping people to question, perhaps by dowsing, and to explore context, we assist others in developing perspective concerning their relationship to a larger whole. We may be able to help them see what is behind some of their problems. This awareness of causation will help them to release these problems, a process which will in itself be healing.

Dowsing is an expression and merging of the ego self in that great sea of the divine heart—the drop in the ocean and the whole ocean emerging as the drop. It establishes our awareness that our actions spring from that sense that we are both part of the whole and the whole. As we become aware of this amazing heritage of the human being, it becomes our responsibility to manifest that awareness to the degree of our understanding of it in all that we do.

Albert Einstein wrote,

> A human being is a part of the whole, called by us the "Universe," a part limited in time and space. He experiences himself, his thoughts and feelings as something separated from the rest— a kind of optical delusion of his consciousness. This delusion is a kind of prison for us, restricting us to our personal desires and to affection for a few persons nearest to us. Our task must be to free ourselves from this prison by widening our circle of compassion to embrace all living creatures and the whole of nature in its beauty [1]

The practicing neophyte dowser now fine-tunes the newly acquired skills in responding to calls and demands, whether in the framework of a job or a personal need. The realization comes that all things are connected; the mineral, plant, animal, and soul kingdoms forming a vast network of relationships which are intertwined, each affecting the other. This creates a feeling of self worth and a responsibility to those being served.

With these higher dimensions of thought, many dowsers become what might be called apprentices, dedicated to the development, continued study, and the art of dowsing. Visions on all levels- the physical, emotional, mental and spiritual- become more perceptive and the inner life more powerful and peaceful. Ideas—gifts from the Intuition—enlighten the mind with ways of helping and healing. The individual no longer feels isolated or separate, but feels a part of the Greater Whole. There is a sense of freedom, expansion, and communication with all other Souls and the universe. The dowser, we might say, becomes a co-creator in the plan, beginning to build new thought forms of goodwill, love and understanding, generating these energies throughout the planetary network, thereby transforming and becoming transformed in the process. We can see that throughout the different stages of dowsing we can greatly affect the planet and all that live therein. Although the idea of interconnection is certainly not new and has been a part of the spiritual teachings of all the great traditions, it is a concept that seems periodically to need to be rediscovered. We have had many warnings in the past thirty years of the ecological folly of our applications of pesticides and disposal of sewage and industrial chemicals, polluting the water and poisoning the land and causing the infusion of heavy metals, pesticides, and other harmful chemicals into the food chain. Now is the time when we simply have to wake up to our responsibility. It is easy to avoid the issues by saying that it is a superhuman job to clean up the waste, the pollution, and the appalling things that we've done to our cities, mountains, rivers, streams, and even our oceans themselves. I'm sure, however, that given sufficient knowledge and an open heart, we can find ways to help.

Geomancy is the process of bringing man's activities into harmony and resonance with the earth and with the cosmos. How much do we need to know about this before we act? The web of relationship is probably much more intricate than we can understand, and so we ask permission of the Source and ask that anything we do may be harmonious and within the balance of nature. There are many practical things that can be done, but the most profound changes come from helping to change consciousness. But because we can do something does not mean that it is appropriate to do so or to the highest good to do it.

Always ask:

May I?— Am I permitted?

Can I? — Am I able? Do I have the skill to do this?

Should I? — Is this the time? Am I the one to do it?

And as a final check, "Is this the truth?"

If, for example, we cause to be moved a harmful-to-health water vein without permission, what neighbors do we affect, what other water supplies? What trees needing water will be neglected; how do we affect the lightning pattern, since it appears that lightning strikes particularly on the crossing point of water veins? Our actions affect the greater whole of existence—and it is our mandate as responsible dowsers to be aware and compassionate.

Our dowsing can have far reaching implications. As we realize our part in the greater whole, new worlds of awareness open for us. We look beyond the five physical senses and see new potentials within ourselves for our personal benefit, for the benefit of others, and for the benefit of our planet.

Can you really be a good dowser without possessing some level of planetary consciousness? I do not think so. How can you do distant or map dowsing unless you have some kind of planetary consciousness? Dowsing encompasses a wholistic philosophy— a philosophy of oneness. That is also what planetary consciousness is about. It is about the integrating of the ego into the heart of universal consciousness. It is the perceiving—through an awareness of the whole—of the right relationship of the parts to the whole, and assisting in creating that harmony. Striving toward planetary consciousness means working toward expansion of your own consciousness, releasing it from the prison of separation, and letting it become limitless.

As you start to develop a planetary consciousness, you will become a better dowser, and you will become a custodian of your own planet Earth, who at this time in history, is very much in need of your help and awareness.

Dowsing is an intuitive art—a means to externalize your own Intuition. The Intuition is a bridge between the conscious self and the Infinite Intelligence, Wisdom, the life force of all creation. By learning to use your Intuition, you can have a positive and beneficial impact on yourself, on others, and on the planet on which we depend for our survival. I have attempted to impart some tools and some practical wisdom and truths to assist you in your day-to-day living. If used properly, the Intuition, and it's tools, will teach you how to improve the quality of your life ten-fold.

By knowing your intuitive mind, you can come to know and understand yourself better in the realms of existence that span the physical, mental, emotional, and spiritual realities. By knowing your intuitive mind, you can relate more fully to the source of creation. Practice the techniques in this book and let the tools, and your Intuition, teach you what is for your highest and greatest good for you in all that you do, so that you may be of benefit to humanity and life in all its forms.

Get Intuit, If In Doubt, Check It Out.

To Write to the Author

We cannot guarantee that every letter written to the author can be answered, but all will be forwarded. Both the author and the publisher appreciate hearing from readers, hoping for your enjoyment and personal benefit from this book. The author sometimes participates in seminars and workshops, and dates and places are announced in the Crystalline quarterly bulletins. To write to the author with your intuitive experiences, to ask a question, or to be put on the bulletin mailing list, write to:

> Dale Olson
> c/o Crystalline Publications
> P.O. Box 2088
> Eugene, Or 97402, U.S.A.

Printed with soy based ink as our part in planetary consciousness.

Recycled Paper.

Bibliography

Archdale, F.A., *Elementary Radiesthesia, and the Use of the Pendulum.* The British Society of Dowsers, 1950.

Bachler, Kathe, *Erfahrungen einer Rutengangerin*, Veritas Verlag, Linz, Harrachstrasse 5, W. Gmy., 1981. (Translated by Marianne Gerhart: *Discoveries of a Dowser*).

Bachler, Kathe, *Discoveries of a Dowser,* (English translation), p.204.)

Beasse, Pierre, *A New and Rational Treatise of Dowsing according to the methods of Physical Radiethesia.* Mokelumne Hill, Ca.:Health Research, 1975.

Bentov, Itzhak, *Stalking The Wild Pendulum*. N.Y.:E.P. Dutton, 1977.

Besant, A., & Leadbeater, C.W., *Thought-Forms.* Il.: Theosophical Publishing House, 1971.

Bhattacharyya, Benoytosh, *The Science of Cosmic Ray Therapy or Teletherapy.* Calcutta, India:Firma KLM Private LTD, 1976.

Bird, C.,&Tompkins, P., *The Secret Life Of Plants.* N.Y.:Avon Books, 1973.

Bird, Christopher, "Dowsing in the United States of America: History, Past Achievements and Current Research." ***The American Dowser***, vol 13, no.3, August, 1973, pp. 105-6.

Bird, Christopher, "Finding It by Dowsing," ***Psychic,*** Vol.VI, no.4, September/October, 1975. pp. 8-13.

Bird, Christopher, ***The Divining Hand***. E.P. Dutton, N.Y., 1979, p.270.

Blackburn, Gabriele, ***The Science and Art of the Pendulum: A Complete Course in Radiethesia.*** Idylwild Books, 1983.

Brennan, Barbara Ann, Hands Of Light: ***A Guide to Healing Through the Human Energy Field.*** Bantam Books, 1987.

Cameron, Verne L., ***Map Dowsing.*** Elsinore, CaEl Cariso Publications, 1971.

Chapman, E.C., ***The 12 Tissue Salts.*** Jove Publications, 1979.

Cox, Bill, **The Techniques of Pendulum Dowsing.** Forces, 1977.

Davis Ph.D., Albert Roy, ***The Anatomy of Biomagnetism.*** Fl.:Davis Research.

Detzler, Robert E., ***Your Mind Net,*** Redmond, Wa., 1988.

Finch, W.J., ***The Pendulum & Possession.*** AZ.:Esoteric Publications, 1975.

Graves, Tom, ***Dowsing and Archaeology***. Turnstone Press, 1980.

Greene, Linda L., ***Learning the Secret Language,*** Samaitan Foundation, Edmond, Ok, 1990.

Judith, Anodea, ***Wheels of Life.*** MN:Liewellyn Publications, 1987.

Kopp, Dr. Joseph, *Transactions of the Swiss Society of Science*, Ebikon, Switzerland, 1970. pp.253-255.

Kopp, Dr. Joseph, "Children's Illneses Due to Soil Influences", *Prophylaxe*, Central Pamphlet For Social Hygiene, Ebikon, Switzerland, Sept. 1970, Vol.9.

Leadbeater, C.W., *The Chakras.* London: Theosophical Publishing House, 1974.

Mermet, Abbe', *Principles and Practice of Radiesthesia.* London:Watkins Publishers, 1935.

Nielsen, Greg & Polansky, Joseph, *Pendulum Power*. Excalibur Books, 1982.

Nadel, Laurie, *Six Sense,* The Whole Brain Book of Intuition, Prentice Hall, New York, 1990

Oregonian, "Are Electrical Lines Powering Cancer?", Kurt Sternlof, August 6, 1987.

Purce, Jill, *The Mystic Spiral,* Thames & Hudson, 1974.

Ramacharaka, Yogi, *Science of Breath.* Yogi Publication Society, 1904.

Raphaell, Katrina, *Crystal Enlightenment.* Aurora Press, 1986.

Raphaell, Katrina, *Crystal Healing.* Aurora Press, 1987.

Ross, T. Edward and Wright, Richard D., *The Divining Mind,* Destiny Books, Vermont, 1990.

The American Dowser, Vermont: American Society of Dowsers, 1960-1975

The American Dowser, Spring 1988 Volume 28, No.2, Herbert Douglas, p24.

The American Dowser, Volume 28, No.4, 1988, The Resonance Factor, T. Edward Ross,II, pg. 9

The American Dowser, 1991, Volume 31, Dowse A House, Richard Perrott, No.1, p.16

The American Dowser, Winter 1991, Vol.31 No.1, Understanding and Dealing With Dowsing Failures, Steven G. Herbert.

Vogel, Marcel, *Psychic Research Newsletter.* Ca.:Psychic Research, 1974-1990.

Wayland, Bruce & Shirley, *Steps to Dowsing Power.* Life Force Press, Inc., 1976.

Quality Living in the 90's

**Develop accurate decision-making skills
Practice Intuitive problem-solving
Experience increased personal power**

In this landmark book, Dale W. Olson takes the reader on an encouraging and inspirational journey using the Intuitive mind for whole-hearted living. He demonstrates straightforward and measurable techniques with various tools, exercises and skills to externalize the Intuition. Most of all, this process gives the readers the confidence to be free from a "trial-and-error", or guessing, approach to life.

Comments From Readers

"Reading your book gave me a deep inner smile and I know there is nothing "out there" that I cannot resolve in perfect harmony."
Rev Johanna K. — Retired minister & teacher, OR

"Your book gave me the confidence and trust to know my Intuition"
David S. — Airline Management, NV

"I'm thankful for your book...showing me how to tap into my higher power. It's great."
Donna B. — LPN, WA

"I consider his service in giving this information, which can add much clarity and balance to our lives, a great kindness."
Aymalee — The Light Magazine, WA

"Inspiring to those of us who desire to raise our quality of living by using all aspects of our being."
Hannah Donovan — Journalist, OR

Your **New Age** library is not complete without this dynamic new book.

KNOWING Your Intuitive Mind by Dale W. Olson • ISBN #1-879-246-00-7 • $12.95
Crystalline Publications • P.O. Box 2088 • Eugene, OR 97402 • (503) 683-8418

Books & Tools for the Intuition

KNOWING Your Intuitive Mind

by Dale W. Olson
ISBN #1-879-246-00-7 $12.95   Quality paperback
196 pages   5½" x 8½"

> Empowering guidebook presenting numerous tools and techniques to develop the Intuition for everyday decision and choice-making.

Knowing Your Intuitive Mind
Pendulum Charts

by Dale W. Olson
ISBN #1-879-246-02-3 $9.95
Cover stock charts (31)

> 31 Charts. Use the Charts in combination with your pendulum to:
>
> ***Choose:*** the proper foods, diet, supplements, careers, jobs, house and locations.
>
> ***Find:*** water, minerals, underground or hidden electrical & water lines, and lost objects.
>
> ***Discover:*** relationship compatibility, personal motivators, and "where to's" from here.
>
> ***Determine:*** Probability of success for any situation, making accurate business decisions, best consumer choices, problem solving for cars, houses, animals, plants, or people.

The uses are unlimited!

Crystal Pendulums

Natural Quartz Crystal with Copper Setting

Large Pendulum	(1¼" +)	$10.95
Small Pendulum	(1" -)	$8.95

Order Direct

Crystalline Publications

P.O. Box 2088 • Eugene, OR 97402
Information or catalog: (503) 683-8418
Orders MC/Visa: 1-800-688-8418

Shipping & Handling	
$0-15.00	add $2.00
$16.00-25.00	add $3.00
$26.00-40.00	add $4.00
$41.00-60.00	add $5.00
Overseas add $10.00 - AIR